NUCLEOTIDE METABOLISM

AN INTRODUCTION

Nucleotide Metabolism

AN INTRODUCTION

J. Frank Henderson

and

A. R. P. Paterson

Cancer Research Unit and Department of Biochemistry
University of Alberta
Edmonton, Alberta, Canada

ACADEMIC PRESS   1973   New York and London

A Subsidiary of Harcourt Brace Jovanovich, Publishers

ACADEMIC PRESS, INC.
111 Fifth Avenue, New York, New York 10003

United Kingdom Edition published by
ACADEMIC PRESS, INC. (LONDON) LTD.
24/28 Oval Road, London NW1

LIBRARY OF CONGRESS CATALOG CARD NUMBER: 72-9996

PRINTED IN THE UNITED STATES OF AMERICA

CONTENTS

PREFACE

The chemistry and metabolism of purines, pyrimidines, and their nucleosides and nucleotides constitute one of the oldest subjects of biochemistry, beginning as it does with the identification of uric acid in 1776. It is ironic that it has taken longer to work out the pathways of the synthesis, interconversion, and catabolism of these compounds than those of many other metabolites.

It is also interesting that nucleotide metabolism has seldom been studied for its own sake. The roles of nucleotides as precursors of nucleic acids and as products of their degradation have prompted many studies, which have usually been quite independent, for example, from investigations of the roles of nucleotides in energy metabolism and in other coenzyme or group-transfer reactions. Another approach has related purine and pyrimidine nucleotide metabolism to human diseases such as gout, orotic aciduria, and the Lesch–Nyhan syndrome. Finally, a large amount of literature has accumulated in connection with the study of the pharmacological uses of antimetabolites of purines, pyrimidines, folic acid, and certain amino acids, all of which at least potentially interfere with one aspect or another of nucleotide metabolism. Most reviews or monographs on these subjects contain sections of varying length on nucleotide metabolism, but no full-length treatment of the subject as a whole has previously appeared.

Although this book appears to be the first devoted solely to nucleotide metabolism, it should be made very clear that it is a textbook rather than an exhaustive monograph, and is selective in scope. It is based on a graduate course given by the authors at the University of Alberta, and attempts to present both a broad view of the field together with more detailed glimpses of a few selected topics which are of particular importance at the present time or of special interest to the authors. Although the historical development of several topics is given briefly, the emphasis is on current developments and the present status of the field.

The references given are also selective, and we ask the indulgence of those whose contributions are not referred to explicitly. Reviews are quoted when possible and appropriate as a source of specific references which they cover, and in other cases the reader is directed to current papers for literature surveys. Usually, therefore, only primary references which are of particular importance or which cannot be found in the secondary sources suggested are listed.

We wish to express our sincere appreciation to Dr. Anita A. Letter for considerable assistance throughout the preparation of the manuscript, and to Miss Daena Letourneau and Mrs. Rosalie Vilonyay for accurately and patiently typing the manuscript and many drafts. We also wish to thank Drs. D. W. S. Westlake, M. K. Robins, and L. B. Brox for reading part or all of the manuscript and for most helpful criticism and suggestions. Finally we wish to thank the authors and publishers, who have kindly permitted the reproduction of original tables and figures, and our students, who over a period of years both stimulated us to prepare this version and have provided critical comments regarding its content and format.

J. FRANK HENDERSON
A. R. P. PATERSON

NOMENCLATURE AND ABBREVIATIONS

Enzymes

The 1964 Recommendations of the International Union of Biochemistry are in general followed for the names of enzymes. The "Recommended Trivial Names" are used in the text. At the ends of most chapters tables give the trivial name, systematic name, and Enzyme Commission number of the principal enzymes discussed. The commonly accepted names of those enzymes not yet listed systematically by the Enzyme Commission have been used.

Purines and Pyrimidines

The structures and numbering systems of the common naturally occurring purine and pyrimidine bases are shown as follows:

Purine and Pyrimidine Bases

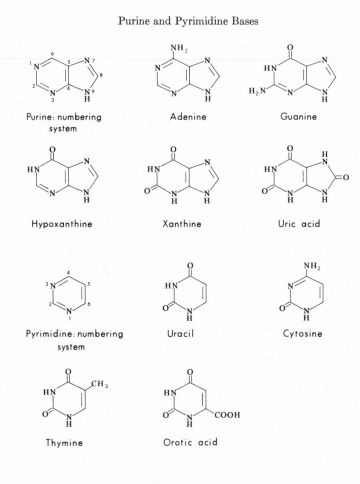

Purine: numbering
system

Adenine

Guanine

Hypoxanthine

Xanthine

Uric acid

Pyrimidine: numbering
system

Uracil

Cytosine

Thymine

Orotic acid

Nucleosides and Nucleotides

The terms "nucleoside" and "nucleotide" in the strictest sense refer to N-glycosides and phosphorylated N-glycosides, respectively, derived from nucleic acids. The term is now used, however, in several broader ways. Thus, adenosine triphosphate (ATP) is not derived from nucleic acids, but is quite legitimately a nucleotide through its relation to adenosine monophosphate (AMP), which is so derived. Other N-ribosides, such as nicotinamide mononucleotide (NMN), are called nucleotides only by extension and analogy, and nicotinamide-adenine dinucleotide (NAD^+), nicotinamide-adenine dinucleotide phosphate ($NADP^+$), etc., are called dinucleotides only by a similar process. Flavin mononucleotide (FMN) is a step still further removed, as it contains ribitol instead of ribose, and flavin-adenine dinucleotide similarly extends the meaning of dinucleotide. N-glycosides such as orotidylate (OMP) and adenylosuccinate are called nucleotides through their close relationship to the "true" nucleotides.

The terms ribonucleoside, deoxyribonucleoside, etc., will be used in preference to riboside, deoxyriboside, etc.

The basic structures of the ribo- and deoxyribonucleosides and nucleotides are as follows:

Ribonucleotides

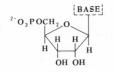

Ribonucleoside 5'- monophosphates

Base: Adenine
       Guanine
       Hypoxanthine
       Succinyladenine
       Xanthine
       Uracil
       Cytosine
       Orotate

Deoxyribonucleotides

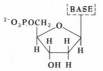

Deoxyribonucleoside 5'- monophosphates

Base: Adenine
       Guanine
       Thymine
       Cytosine
       Uracil

Intermediates of Purine Biosynthesis *de Novo*

The intermediates of the pathway of purine biosynthesis *de novo*, particularly those which are acyclic, are only called nucleotides by a further extension of the original definition. Although these names are used by tradition, the Enzyme Commission has preferred to regard these compounds as simple phosphoribosyl derivatives. Both old and newer terms are given here; only names based on the Enzyme Commission usage will be used in the text, and they are not abbreviated.

Old Name	Name Used Here
Phosphoribosylamine (PRA)	Ribosylamine phosphate
Glycineamide ribonucleotide (GAR)	Phosphoribosyl glycineamide
Formylglycineamide ribonucleotide (FGAR)	Phosphoribosyl formylglycineamide
Formylglycineamidine ribonucleotide (FGAM)	Phosphoribosyl formylglycineamidine
Aminoimidazole ribonucleotide (AIR)	Phosphoribosyl aminoimidazole
Aminoimidazolecarboxylate ribonucleotide (CAIR)	Phosphoribosyl aminoimidazole carboxylate
Aminoimidazole succinocarboxamide ribonucleotide (SAICAR)	Phosphoribosyl aminoimidazole succinocarboxamide
Aminoimidazolecarboxamide ribonucleotide (AICAR)	Phosphoribosyl aminoimidazole carboxamide
Formamidoimidazolecarboxamide ribonucleotide (FAICAR)	Phosphoribosyl formamidoimidazole carboxamide

Abbreviations

The IUPAC-IUB Combined Commission on Biochemical Nomenclature has promulgated two sets of abbreviations and symbols for nucleotides and related compounds. While these are suited for description of polynucleotides, the distinction between bases and nucleosides is not always immediately obvious and this distracts from their pedagogical use. The abbreviations used here are, we feel, intuitively obvious and well suited to the portrayal of reaction schemes in which the addition or subtraction of groups occurs.

Name	Abbreviation
Phosphate ion (PO_4^{3-})	P_i
Phosphoryl group (PO_3^{2-})	Ⓟor P
Pyrophosphate ion	PP_i
Pyrophosphoryl group	PP
Ribosyl group	R
2-Deoxyribosyl group	dR
BASES	
Adenine	A
Guanine	G

BASES (*continued*)

Hypoxanthine	H
Xanthine	X
Uracil	U
Cytosine	C
5-Hydroxymethylcytosine	HM-C
Thymine	T
Orotic acid	O
Any purine base	Pu
Any pyrimidine base	Py
Any base	B

Ribonucleosides and 2'-Deoxyribonucleosides

Abbreviations for ribonucleosides and 2'-deoxyribonucleosides are derived from those for the bases plus those for the ribosyl or 2'-deoxyribosyl groups. Thus AR stands for adenosine, UR for uridine, and BR for any ribonucleoside; AdR stands for deoxyadenosine, UdR for deoxyuridine, and BdR for any deoxyribonucleoside. When used in abbreviations for nucleotides, N also stands for any ribonucleoside and dN for any deoxyribonucleoside.

Ribonucleoside and 2'-Deoxyribonucleoside Monophosphates

In most cases the traditional abbreviations based on the term "nucleoside monophosphate" are used. Thus AMP stands for adenosine monophosphates (adenylate), UMP for uridine monophosphate (uridylate), and NMP for any ribonucleoside monophosphate. Similarly, dAMP stands for deoxyadenosine monophosphate (deoxyadenylate), dUMP for deoxyuridine monophosphate (deoxyuridylate), and dNMP for any deoxyribonucleoside monophosphate. To illustrate certain points, the base, sugar, and phosphate group are indicated individually: AMP = ARP; dUMP = UdRP; etc.

Nucleoside Di- and Triphosphates

In most cases the traditional abbreviations based on the terms "nucleoside diphosphate" (NDP) and "nucleoside triphosphate" (NTP) are used. Thus ADP stands for adenosine diphosphate and dATP for deoxyadenosine triphosphate. To illustrate certain points, the base, sugar and phosphate groups are indicated individually: ADP = ARPP; dATP = AdRPPP, etc.

Other Terms

Deoxyribonucleic acid	DNA
Ribonucleic acid	RNA
Coenzyme A	CoA
Flavin-adenine dinucleotide	FAD
Riboflavin 5′-phosphate	FMN
Nicotinamide-adenine dinucleotide	NAD^+
Nicotinamide-adenine dinucleotide phosphate	$NADP^+$
Nicotinamide mononucleotide	NMN
Tetrahydrofolate	H_4-folate
5′-Deoxyadenosylcobalamin	Coenzyme B_{12}

GENERAL ASPECTS OF NUCLEOTIDE METABOLISM

The discovery of a large number of nucleotides as normal tissue constituents, and their separation, isolation, and characterization (Chapter 1), their structures in solution (Chapter 2), and a general discussion of the many and varied functions of nucleotides in cells (Chapter 3), provide a background and general framework for the discussion of the synthesis, interconversion, and catabolism of these natural products which follows in later chapters.

Certain group-transfer reactions are common to many aspects of nucleotide metabolism. These include phosphoryl group transfers which generate the nucleoside polyphosphates found in the cells (Chapter 4), and one-carbon and amino group transfers by which purine and pyrimidine rings are formed and interconverted (Chapter 5). Finally, the formation of the ribose moiety of nucleotides is considered (Chapter 6) to conclude Part I.

CHAPTER 1

NUCLEOTIDES AS TISSUE CONSTITUENTS

I. Introduction

The term *nucleotide* was introduced in 1908 by Levene to refer to phosphate esters of *nucleosides,* which were so named because they contained a sugar in glyco*sidic* linkage with the purine and pyrimidine bases of the *nucleic* acids. The latter term, of course, represents the source from which the bases were first isolated, the *nuclei* of animal cells.

The purines and pyrimidines are major chemical constituents of cells and occur primarily as components of polymerized nucleotides (nucleic acids) and to a much lesser extent in the form of "free" (that is, unassociated) nucleotides. Free nucleosides and bases usually represent a very small fraction of the total purine and pyrimidine content of living cells. However, there are exceptions to this generalization, such as the occur-

rence of substantial amounts of theophylline, theobromine, and caffeine
in some plant tissues, and the occurrence of the arabinoside of thymine
and uracil in the Caribbean sponge, *Cryptotethya crypta*. Guanine, as the
silvery constituent of fish scales, and the excretion product, uric acid, are
free purine products of animal cells, but only occur in substantial amounts
extracellularly.

The free nucleotides of cells and tissues are usually isolated from aqueous
extracts, the preparation of which involves the precipitation of proteins
and nucleic acids with 70% alcohol or with acids (perchloric or trichloroacetic acids are commonly used). The resulting "acid-soluble" fraction
of cells and tissues contains low molecular weight, water-soluble components of the cell, including a complex mixture of free nucleoside phosphates. Modern chromatographic methods with ion-exchange media are
able to resolve this mixture, the complexity of which is illustrated in Fig.
1-1; this figure represents the resolution of the mixture of free nucleotides
in an acid-soluble extract of mouse liver by chromatography on a column
of DEAE-Sephadex. The strong absorption of ultraviolet light in the 250–
300 mμ region of the spectrum exhibited by purines, pyrimidines, and

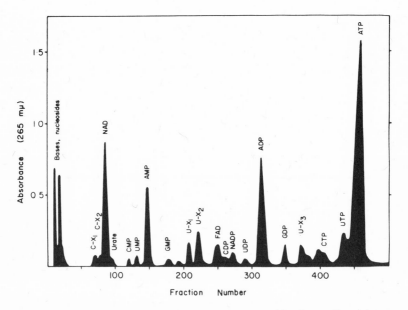

FIG. 1-1. Acid-soluble nucleotides of mouse liver. Rapidly frozen liver tissue was
extracted with 0.4 M perchloric acid; the extract was neutralized, chromatographed on
a column of DEAE-Sephadex, and the absorbance of eluate fractions at 265 mμ was
measured. From (*2*). Reproduced with permission.

their nucleotide derivatives is used in their quantitation (*1*); this facile and sensitive analytical method has been a critical element in the remarkable expansion of nucleic acid and nucleotide biochemistry in recent times. It is seen in Fig. 1-1 (*2*) that the mono-, di-, and triphosphates of the ribonucleosides represented in RNA were present in the soluble fraction of liver. The free nucleotide composition of tissues is actually more complex than here indicated because minor nucleotide components, such as the deoxyribonucleoside phosphates, are also present, but remain undetected without the use of special methods. Bases and nucleosides (the two leading peaks) are present, but only in small quantities.

II. Historical Development

Our knowledge of the free nucleotides of tissues has developed in step with our understanding of the chemical nature and metabolism of the nucleic acids. A useful perspective may be afforded by the following brief resumé of the historical development of the nucleotide field.

A. THE EARLY PERIOD

The nucleic acids were discovered by Miescher in 1868–1869, when he isolated from pus cell nuclei a material which contained phosphorus, was soluble in alkali, but precipitated under acidic conditions. This material was subsequently prepared from other sources and when freed from protein it was called "nucleic acid," a term introduced by Altman in 1889. The classical preparations of nucleic acid from yeast yielded a product which we now recognize as ribonucleic acid (RNA). The nucleic acid prepared from thymus glands, "thymonucleic acid," was also extensively studied; this material [which, in present terms, was deoxyribonucleic acid (DNA)] was different from yeast nucleic acid. From hydrolysates of these preparations the heterocyclic bases were isolated and characterized. At one time, yeast and thymus nucleic acids were thought to be representative of plant and animal nucleic acids, respectively (*3*). By 1909, it was apparent that yeast nucleic acid contained adenine, guanine, cytosine, uracil, phosphoric acid, and a sugar which Levene showed at that time to be D-ribose. Thymonucleic acid yielded adenine, guanine, cytosine, thymine, phosphoric acid, and a sugar which was not identified correctly until 1929, when it was characterized as 2-deoxy-D-ribose.

By 1912, ribonucleosides representing the four bases of yeast nucleic acid had been isolated from hydrolysates and their chemical structures determined correctly, except for the improper assignment of the glycosidic linkage to purine N-7. However, chemical hydrolysis did not yield nucleosides from thymonucleic acid and this problem remained unsolved until enzymatic hydrolysis was employed in 1929. Thus, phosphate esters of nucleosides, i.e., nucleotides, were indicated as structural units of nucleic acids. The term "nucleotide" had been introduced in 1908 by Levene in reference to a hydrolysis product of thymonucleic acid, which he had named thymidylic acid. This compound was recognized as a phosphoric ester of a thymine nucleoside (4). By 1918, the four nucleotide subunits of RNA had been isolated in pure form from alkaline hydrolysates of yeast nucleic acid; however, the position of the phosphate residue was uncertain at that time.

In 1925, the structure of inosinic acid, which was thought to be the 5′-monophosphate of inosine, was confirmed. Inosinic acid, the first free nucleotide to be recognized, has an interesting history (4). In 1847, the barium salt of this substance was isolated from beef extract by Leibig, who derived the name from the Greek words for muscle fiber. The presence of phosphorus in this substance was not recognized until 1895. In 1909, Levene determined the structure of inosine, a hydrolysis product of inosinic acid, to be the riboside of hypoxanthine; with this information, the nucleotide structure of inosinic acid became apparent.

An important discovery, that of free adenylic acid in muscle, was made by Embden in 1927. Muscle adenylate was recognized as the 5′-monophosphoric ester of adenosine because enzymatic deamination yielded the known inosinic acid. It was shown at that time that the deaminase preparations from muscle did not deaminate the adenylic acid isolated from alkaline hydrolysates of yeast nucleic acid; as well, differences were apparent in the chemical properties of the adenylic acids from these two sources. Yeast adenylic acid and the other nucleotides from alkaline hydrolysates of RNA were ultimately shown to be mixtures of the 2′- and 3′-phospho esters. In 1929 the isolation of adenosine triphosphate from muscle was reported by Lohmann and independently by Fiske and Subbarow. The discovery of adenosine diphosphate followed in 1935.

The lability of the sugar component of thymonucleic acid had frustrated attempts to isolate structural subunits by chemical hydrolysis; however, in 1929 a gentle hydrolytic procedure using dog intestinal enzymes yielded the expected four nucleosides, and the sugar was then characterized as 2-deoxy-D-ribose. Thus, it was evident that both types of nucleic acid were polymers of nucleotides. After recognition of DNA in plant tissues and the demonstration of RNA in animal tissues, it was apparent that cells in general contained both types of nucleic acid.

By analogy with muscle adenylate, Levene and Bass (4) anticipated the discovery in tissues of other free nucleotides related to the nucleic acid components. Such nucleotides were indeed found, but their discovery had to await the advent of an adequate technology, which did not become available until after 1949. Prior to the period of recent development in the nucleotide area, several nucleotide coenzymes were discovered:

Nicotinamide adenine dinucleotide phosphate, Warburg *et al.* (1935)
Nicotinamide adenine dinucleotide, Schlenk and von Euler (1936)
Flavin adenine dinucleotide, Warburg and Christian (1938)
Coenzyme A, Lipmann (1946)

Prior to 1936, the purine nucleotides were formulated as having the pentose at N-7 because of an arbitrary assignment to this position by Fischer. In 1936, Gulland was able to assign the glycosidic bond to N-9 on the basis of comparisons of absorption spectra of synthetic N-7 and N-9 substituted purines. During the 1940's, the chemistry of the nucleosides was developed extensively and the furanose nature of the sugar was established, as was the β-configuration of the glycosyl linkages (*5*).

B. The Post-1949 Period

During the 1940's, a new technology was developed for the manipulation and quantitation of microgram quantities of the nucleic acid bases. The advent of this technology, which involved paper chromatography, chromatography on ion-exchange media, and ultraviolet spectrophotometry, brought about a literal explosion in our knowledge of the free nucleotides of cells and tissues.

In 1949, Cohn applied ion-exchange chromatography to the separation of nucleic acid derivatives and demonstrated elegant separations of nucleotides on anion-exchange resins such as Dowex-1 (which has quaternary ammonium charged groups on a polystyrene matrix). It was shown at that time that both RNA and DNA could be hydrolyzed by phosphodiesterases to yield nucleoside 5'-phosphates and it was therefore evident that polynucleotides could be regarded as assemblies of nucleoside 5'-phosphates. At this point, the adenosine phosphates were the only free nucleotides known to occur in tissues, apart from the coenzymes. Because the free adenosine phosphates were 5'-esters, a precursor relationship between these compounds and the polynucleotides seemed likely.

A major step was taken by V. R. Potter and his associates in experiments which, for the first time, applied anion-exchange chromatography to the isolation of nucleotides from the acid-soluble fractions of liver and other tissues. A complex mixture of free nucleotides was at once shown to be present in tissues (see Fig. 1-2). The 5'-mono-, di-, and triphosphates of

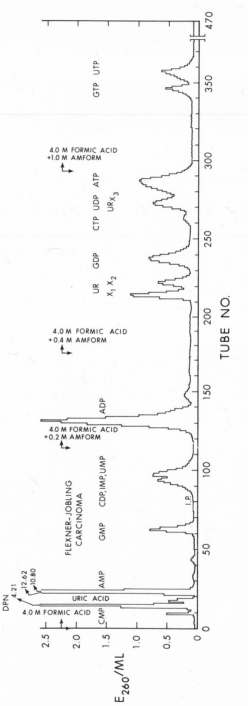

Fig. 1-2. Acid-soluble nucleotides of the Flexner-Jobling rat carcinoma. Pooled tumor tissue was extracted with 0.4 *M* perchloric acid; the neutralized extract was chromatographed on a column of Dowex-1 resin (formate form) using gradient elution with formic acid–ammonium formate solutions. From (8). Reproduced with permission.

the four ribonucleosides present in RNA were identified, as were several nucleotide conjugates (6–8). Deoxyribonucleoside phosphates were not evident in these studies, but were found subsequently by others.

Free deoxyribonucleotides are evidently formed in cells at the time of DNA synthesis (9) and are not present in appreciable quantities in non-proliferating cells. By rechromatographing nucleotide-containing fractions isolated from calf thymus extracts by anion-exchange chromatography, R. L. Potter and co-workers (10) were able to isolate the mono-, di-, and triphosphates of deoxycytidine and thymidine. dATP was isolated from a transplantable rat tumor by LePage (11). It became evident that only very small quantities of free deoxyribonucleosidic materials were present in animal tissues; several indirect methods of high sensitivity have been developed to meet this analytical problem. Deoxyribonucleosidic substances in tissue extracts have been measured by highly sensitive microbial assays using lactobacilli with growth requirements for deoxyribonucleosides (12, 13). Pool sizes of the deoxyribonucleotides have also been measured by an indirect method in which cells were cultured in media containing ^{32}P-phosphate of known specific activity; the acid-soluble nucleotides extracted from such cells were determined by means of the radioactivity which co-chromatographed with samples of authentic deoxyribonucleotides (14). By another indirect method employing DNA polymerase, pool sizes of dATP, dGTP, dCTP, and dTTP have been measured in cultures of mouse embryo cells (15, 16).

With the advent of an adequate nucleotide technology and the development of widespread interest in nucleotide metabolism, many free nucleotides have been found in cells; however, no comprehensive survey has appeared since the 1958 report of Henderson and LePage (17), which lists over 100 nucleotides, and the list has probably more than doubled since. A large number of free nucleoside diphosphate conjugates have been isolated from cells and characterized; many such compounds serve as coenzymes in group-transfer reactions, for example, CDP-X and UDP-X derivatives participate, respectively, in phospholipid and polysaccharide synthesis (see Chapter 3). As well, many new nucleotides have been recognized as metabolites of nucleoside antibiotics and of synthetic purine and pyrimidine analogues.

III. Free Nucleotides of Cells and Tissues

A. The Variety of Free Nucleotides in Nature

So many free nucleotides are known that a current, comprehensive listing of them becomes a major task beyond the scope of this text; various past reviews may be consulted (5, 17–24). The following discussion aims at providing the reader with some idea of the variety of nucleosidic and nucleotidic compounds that are found in the unpolymerized form in cells. The *ribo*nucleotides are the most abundant and varied species in nature. Although *deoxy*ribonucleotides in the polymeric form, DNA, constitute a major tissue constituent, free deoxyribosidic compounds are found only in small concentrations and for the most part only as simple nucleotides. In animal cells, a few nucleotides with sugars other than ribose and deoxyribose are known.

1. Ribonucleotides

a. The Phosphates and Polyphosphates of the RNA Ribonucleosides. The 5′-mono-, di-, and triphosphates of adenosine, guanosine, uridine, and cytidine, the four ribonucleosides found in RNA, represent the bulk of the cellular ribonucleotides and occur in virtually all cells. The triphosphates are the immediate precursors of the polyribonucleotides and, as well, are involved in energy metabolism, synthetic reactions and other functions outlined in Chapter 3. Transphosphorylations between the three levels of phosphorylation occur with facility and, although the triphosphates are normally the most abundant, the relative amounts of the three phosphates are related to the energy state of the cell.

Higher polyphosphate derivatives of purine ribonucleosides are known. Adenosine tetraphosphate has been isolated from horse muscle; the compound appears to be a 5′-polyphosphate, but the structure of the polyphosphate portion is uncertain and a biological role is unknown (*25*). Guanosine tetraphosphate and adenosine pentaphosphate have also been reported.

b. Cyclic Monophosphates. Adenosine 3′,5′-cyclic monophosphate and the corresponding guanosine derivative have been detected in animal tissues and are also found in urine (*26*). Cyclic AMP mediates the action of a number of hormones and has been termed "an intracellular second messenger" (Chapter 3). Functions of the more recently discovered cyclic GMP have yet to be reported.

c. Nucleotide Biosynthetic Intermediates. A number of ribonucleoside monophosphates, the bases of which are not represented in the nucleic acids, are known to be intermediates in nucleotide biosynthesis *de novo*, nucleotide interconversions, and coenzyme synthesis. The most prominent of these, inosine monophosphate, is found in the acid-soluble fraction of many tissues; others have been recognized only as intermediates in enzymatic reactions and presumably are present in cells only in small amounts. The di- and triphosphates of inosine do not appear in animal cell extracts.

d. *Conjugated Ribonucleotides.* A varied and substantial group of ribonucleotide conjugates occurs in cells; these compounds may be regarded as ribonucleoside monophosphate-X, or ribonucleoside diphosphate-X where X is some group other than phosphate. These compounds have various functions as coenzymes; examples are given in Table 1–I (it should be noted that this listing is not comprehensive).

In Table 1–I the coenzymes involved in electron transport are classified as ribonucleoside diphosphate conjugates. The number of known nucleotides is swelled by the large number of ribonucleoside diphosphate conjugates which are coenzymes or group-carrying intermediates in the synthesis of various substances such as phospholipids, polysaccharides, cell wall substances, branched-chain sugars, and mucopolysaccharides. Certain of these compounds occur in substantial amounts in cell extracts and are prominent in chromatographic nucleotide profiles; for example, peaks attributable to uridine diphosphate sugars are found on anion-exchange chromatograms of liver extracts (see Fig. 1-1, peaks UX_1 and UX_2), of eye lens (*22*), and of lactating mammary gland (*27*). Guanosine diphosphate glucose is also prominent in extracts of the latter tissue. The occurrence of such nucleotides is a reflection of the specialized functions and products of tissues and thus of cell differentiation.

The lists in Table 1–I are intended to illustrate the great variety of the ribonucleoside diphosphate conjugates that occur in cells and tissues; for more extensive treatment and original references, reviews may be consulted (*5, 17–24*).

e. *Nucleosidyl Derivatives.* Several "adenosyl" compounds are known which are 5'-derivatives of 5'-deoxyadenosine: *S*-adenosylmethionine, *S*-adenosylhomocysteine, and *S*-adenosylethionine. In 5'-deoxyadenosylcobalamin, vitamin B_{12} coenzyme, the 5'-deoxyadenosyl group is bound in an unusual linkage through the 5'-methylene carbon to the cobalt atom.

2. Deoxyribonucleotides

As we have already noted, the deoxyribonucleoside phosphates occur in animal tissues in small amounts and their concentrations are increased in cells engaged in DNA synthesis. The triphosphates of the four deoxyribonucleosides represented in DNA have been demonstrated in cell extracts, as have the mono- and diphosphates of thymidine and deoxycytidine. The mono-, di-, and triphosphates of deoxyuridine are known as intermediary compounds in the metabolism of the pyrimidine deoxyribonucleotides.

TABLE 1–I

EXAMPLES OF CONJUGATED NUCLEOTIDES:
COMPOUNDS IN WHICH A NUCLEOTIDE
PHOSPHORYL GROUP IS ESTERIFIED, OR IN
ANHYDRIDE LINKAGE, WITH "X," A GROUP
OTHER THAN PHOSPHATE

A. Ribonucleoside monophosphate-X
    Acyladenylates
    Adenosine 5'-phosphosulfate
    3'-Phosphoadenosine 5'-phosphosulfate
    CMP-N-acetylneuraminic acid
B. Ribonucleoside diphosphate-X
    Adenosine diphosphate-X derivatives
        NAD, NADP, deamido-NAD
        CoA, dephospho-CoA, acyl-CoA derivatives
        FAD
        ADP-hexoses: D-glucose, D-mannose, D-galactose
        ADP-mannitol
    Guanosine diphosphate-X
        GDP-hexoses: D-glucose, D-mannose,
                D-fructose, L-galactose,
                    L-fucose, D-rhamnose
        GDP-D-mannuronic acid
        GDP-4-keto-6-deoxy-D-mannose
        P^1,P^4-Diguanosine tetraphosphate
        P^1,P^3-Diguanosine triphosphate
        Guanosine tetraphosphate
    Cytidine diphosphate-X
        CDP-D-glucose
        CDP-L-ribitol
        CDP-abequose (3,6-dideoxy-D-$xylo$-hexose)
        CDP-ethanolamine
        CDP-choline
        CDP-D-glycerol
    Uridine diphosphate-X
        UDP-hexoses: D-glucose, D-galactose,
                D-fructose, L-rhamnose
        UDP-pentoses: D-xylose, L-arabinose
        UDP-hexuronic acids: D-glucuronic,
                    D-galacturonic,
                    L-iduronic
        UDP-N-acetylhexosamines: D-glucose,
                      D-galactose,
                      D-mannose
        UDP-N-acetylmuramic acid
        UDP-N-acetylmuramyl peptides
        UDP-N-acetylgalactosamine 4-sulfate
C. Thymidine diphosphate-X
    dTDP-hexoses: D-glucose, D-mannose,
                D-galactose, L-rhamnose
    dTDP-ribose
    dTDP-4-keto-6-deoxy-D-glucose

Certain bacterial phages have DNA with unusual pyrimidine constituents; in the phage-infected bacterial cell, phage-coded enzymes effect the synthesis of free deoxyribonucleotides containing the unusual bases, prior to their incorporation into the phage DNA. In the DNA of bacteriophages T2, T4, and T6, 5-hydroxymethyldeoxycytidylate is found in place of deoxycytidylate, and bacteriophage SP8 of *Bacillus subtilis* contains 5-hydroxymethyldeoxyuridylate in place of DNA thymidylate. In both cases, the 5-hydroxymethyl group is introduced during the biosynthesis of the free nucleotide precursors of the phage DNA's and the appropriate 5-hydroxymethyldeoxyribonucleotides are formed at the three levels of phosphorylation.

A number of conjugated deoxyribonucleoside diphosphates are known, although those known are less numerous than similar compounds of the ribose series. For example, dCDP-choline and dCDP-ethanolamine have been found in sea urchin eggs and in animal tissues, but their significance is unknown. Various thymidine diphosphate sugars (see Table 1–I) have been isolated from bacterial cells; these compounds serve as coenzymes in the synthesis of polysaccharides in *Salmonella typhimurium*.

3. Other Nucleotides

The nucleoside antibiotics comprise a varied group of substances which are structurally analogous to the nucleic acid nucleosides. These substances have potent biological activity which results from interference with nucleotide and nucleic acid metabolism in treated cells. The nucleoside antibiotics shown in Figs. 1-3 and 1-4 are, with the exception of decoyinine, converted by cells into nucleotide derivatives. These nucleoside analogues and their metabolites are mentioned in this particular context because they are naturally occurring substances and so represent an aspect of the great variety of nucleotides in nature. There exist also a large number of synthetic analogues of the nucleic acid bases, many of which are metabolized by way of their nucleotide derivatives; some of these are discussed elsewhere in this text. For a fuller consideration of the nucleoside antibiotics and access to the original literature, several reviews may be consulted (*28–30*).

Figure 1-3 shows a number of nucleoside antibiotics which contain adenine and an unusual sugar at the 9-position. Cordycepin (3'-deoxyadenosine) is a cytostatic agent and is converted to the 5'-mono-, di-, and triphosphate derivatives in mouse tumor cells. *Cordyceps militaris*, the organism which produces cordycepin, also produces a related analogue with antitumor activity, 3'-amino-3'-deoxyadenosine; this compound is also metabolized by way of phosphate derivatives. The "angustmycins," psicofuranine and decoyinine, are, respectively, the 9-β-D-psicofuranosyl

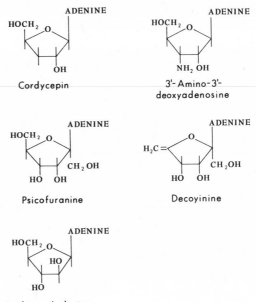

FIG. 1-3. Some nucleoside antibiotics which are 9-glycosyladenines.

and 9-β-D-(5,6-psicofuranoseenyl) derivatives of adenine. Both compounds, as nucleosides, inhibit guanylate synthetase; in addition, psicofuranine is phosphorylated. Arabinosyladenine is cytotoxic for animal cells and also has important antiviral activity; derivatives at the three levels of phosphorylation are formed. Arabinosyladenine was found as a naturally occurring nucleoside produced by a streptomycete after the biological properties of the chemically synthesized product had been recognized. Arabinosylcytosine, a synthetic product not yet found in nature, is similarly metabolized and is an important agent in the chemotherapy of cancer.

The several nucleoside antibiotics shown in Fig. 1-4 are β-D-ribofuranosides with unusual bases; these compounds are phosphorylated in cells and their biological effects are achieved by way of nucleotide metabolites. Tubercidin is one of several 7-deazaadenosine derivatives produced by certain streptomycetes; the related compounds toyocamycin and sangivamycin have cyano and carboxamido groups, respectively, at position-5 of the pyrrolo[2,3-d]pyrimidine ring. These compounds are highly cytotoxic and virocidal. Nebularine is a toxic substance isolated from the poisonous mushroom *Agricus nebularis*. 5-Azacytidine is a potent antileukemic agent; it was also synthesized chemically before its isolation from culture filtrates of a streptomycete.

B. NUCLEOTIDE COMPOSITION OF CELLS AND TISSUES

The nucleotide profiles of Figs. 1-1 and 1-2 show that cells and tissues contain complex mixtures of nucleotides. The principal components of such mixtures can be isolated and determined using current methods of column and thin-layer chromatography with anion exchange media. However, a crucial step prior to analysis is the extraction of nucleotides from cells. With some tissues, the extraction procedure has a profound influence on the analytical result and special precautions must be taken if the result is to truly represent the nucleotide composition of living cells.

1. Preparation of Cell and Tissue Extracts

Free nucleotides are separated from cellular macromolecular substances by cold extraction procedures in which the latter are precipitated; the usual extractants are dilute acids [10% trichloroacetic, or 0.4–0.6 M perchloric acid, and 0.3 M acetic acid have been recommended (31)] or alcohols (70% ethanol or 60% methanol). The manipulations to which the

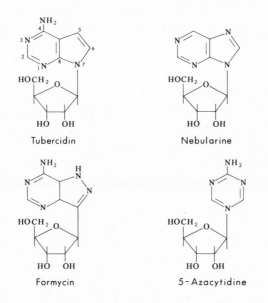

FIG. 1-4. Some ribonucleoside antibiotics with unusual bases.

sample of cells or tissue are subject preliminary to the extraction are of critical importance if the analytical results are to reflect the *in vivo* circumstances. For example, with rat brain, a dissection lasting 20 seconds is sufficient to reduce the ATP content of the tissue by 70% (*22*). The ATP levels and ATP/ADP ratios of rat liver are subject to very rapid and profound changes due to anoxia resulting from interruption of blood circulation during sampling. For example, Chance *et al.* reported that an ATP/ADP ratio of 6.8 in rat liver fell to 1.4 when the portal vein was clamped for 60 seconds (*32*). Rapid freezing methods have been employed to minimize such changes. Liver samples are taken from the tissue *in situ* by rapidly squashing a lobe between two metal blocks attached to tongs; this apparatus is precooled in liquid nitrogen. Tissue that is not compressed and instantly frozen between the blocks is discarded. For the analysis of superficial rodent tumors, the entire animal may be plunged into liquid nitrogen and the tumor dissected while frozen; however, this method does not cause instantaneous freezing. Rapidly frozen tissue samples are pulverized with instruments precooled in liquid nitrogen and, while frozen, tissue powders are dispersed into cold perchloric acid. In simpler procedures, employed when changes in nucleotide patterns do not occur or are unimportant, tissues are simply homogenized in the acidic extractant.

The nucleotide patterns of cultured animal cells and ascites tumor cells are generally less vulnerable than those of tissues; however, if washing is required, a medium capable of supporting energy metabolism should be employed. If cells are to be packed by centrifugation prior to extraction, cultures or ascitic fluids may first have to be cooled. Cells are extracted in the cold; some procedures employ alternate freezing and thawing in the presence of acidic extractants. Extraction of cultured mouse embryo cells with 60% methanol for 16 hours at $-20°C$ has been employed in the analysis of deoxyribonucleoside triphosphates (*9*).

Prior to analysis, perchloric acid extracts are neutralized with KOH ($KClO_4$ precipitates) and those made with trichloroacetic acid are extracted with ether to remove the acid; with alcoholic extracts and those made with formic or acetic acid, the volatile precipitant is removed under vacuum.

2. Nucleotide Analysis

Acid-soluble extracts can be assayed for certain nucleotides by enzymatic methods. Various assay procedures for the adenosine phosphates and the pyridine nucleotide coenzymes are available (*22,33*). Specific methods for the determination of picomole amounts of the four deoxyribonucleoside triphosphates have been reported recently (*15,16*).

Highly developed chromatographic methods employing anion-exchange media with thin-layer, paper, or column techniques are available for the resolution of complex nucleotide mixtures (for example, see Fig. 1-1); such methods have been surveyed by Grav (*34*).

For individual nucleotides isolated by chromatographic procedures, the ultraviolet absorption spectra provide facile identification and quantitation (*1*). Well-known chemical (colorimetric) methods are available to identify and quantitate the sugar component, and to measure the relative amounts of sugar and phosphate in isolated nucleotide fractions (*34*). Enzymatic methods are also used in characterizing and in quantitation of nucleotides (*35*).

3. Nucleotide Concentrations in Particular Tissues

The phosphates of the four ribonucleosides represented in RNA are found in virtually all cells, with erythrocytes being exceptional in that not all four ribonucleosides are represented in some nonnucleated cells. Nucleotide coenzymes and conjugated nucleotides are present in patterns which are characteristic of the tissue sample. These compounds are all nucleoside 5′-phosphate esters and the triphosphates are the most abundant species under conditions adequate for the maintenance of cellular energy metabolism. The nucleotide composition of a variety of animal tissues is discussed by Mandel (*22*); we will consider only several.

The following data about liver are presented to exemplify the nucleotide composition of an animal tissue, but are not intended to be typical of

TABLE 1–II

RIBONUCLEOTIDE CONCENTRATIONS IN
NORMAL RAT LIVER SAMPLED BY
RAPID FREEZING *in Situ*[a]

Nucleotide	Tissue concentration (mμmoles/g fresh wt)
ATP	2580
ADP	725
AMP	106
UTP	330
UDP	45
UMP	72
CTP	180

[a] From (*36*). Reproduced with permission.

TABLE 1–III

NUCLEOTIDE POOLS IN EXPONENTIALLY PROLIFERATING
CHICK EMBRYO FIBROBLASTS IN CULTURE[a,b]

| | Time in culture with ^{32}P-phosphate | |
Nucleotide	2 hours	12 hours
ATP	1890	2390[c]
CTP	53	73
GTP	190	220
UTP	130	180
dATP	12	19
dCTP	7.1	8.4
dGTP	5.3	6.1
dTTP	13	15
AMP	73	130
CMP	10	
UMP	230	270
ADP	390	520
UDP	29	34

[a] From (14). Reprinted from *Biochemistry* 9, 917 (1970). Copyright (1970) by the American Chemical Society. Reprinted by permission of the copyright owner.

[b] Cells were cultured in the presence of ^{32}P-phosphate for the indicated period, washed, and extracted with cold 0.4 M perchloric acid. The extract was neutralized and chromatographed with the nucleotides listed above as carriers in a two-dimensional, thin-layer system which resolved the carrier mixture. The ^{32}P activity which accompanied each carrier spot and the specific activity of the medium photphate were used to calculate pool sizes.

[c] Values are moles $\times 10^{-12}$ per 10^6 cells.

cells and tissues; in fact, of animal tissues, liver has one of the highest nucleotide contents, about 8000 mμmoles of free nucleotide per gram fresh weight (22). The concentrations of various ribonucleoside phosphates in normal rat liver are listed in Table 1–II (36). In contrast, the total free deoxyribosidic material in rat liver has been determined to be about 50 mμmoles per gram fresh weight, and about 50% of this is free deoxycytidine. (37). Free thymidine compounds represent a small proportion of the total content of deoxyribosidic materials in normal rat liver; the total of thymidine and thymidine phosphates is about 4 mμmoles per gram of fresh liver (38).

In regenerating rat liver, the size of the free deoxyribonucleotide pool is increased two- to threefold over that of normal liver (37), whereas that of the ribonucleotides is changed only to a minor extent (36).

The nucleotide pools of chick fibroblasts proliferating exponentially in culture have been determined by an isotopic method; as seen in Table 1–III, in these cells the pool of ATP is about 10 times that of any other ribonucleoside triphosphate and more than 100 times that of dATP or dTTP (14). The small size of the deoxyribonucleotide pool in the exponentially proliferating cells is consistent with observations by others and suggests that the deoxyribonucleotides are synthesized as needed for DNA synthesis and do not accumulate. Skoog and Nordenskjöld (9) have estimated that in cultures of mouse embryo cells undergoing DNA synthesis, the pool of dGTP was sufficient to support only about 30 seconds of DNA

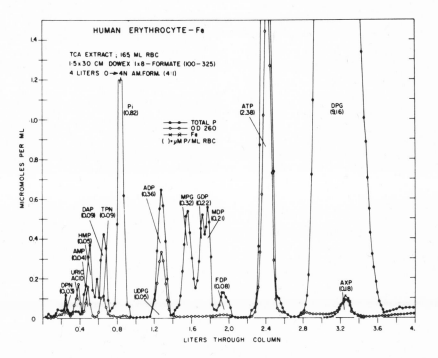

Fig. 1-5. Nucleotides of the human erythrocyte. Trichloroacetic acid extracts of fresh erythrocytes were chromatographed on a column of Dowex-1 resin (formate form) eluted with a gradient of ammonium formate. [From G. Bartlett, Patterns of phosphate compounds in red blood cells of man and animals, in *Advances in Experimental Medicine and Biology* (G. J. Brewer, ed.), Vol. 6, p. 245. Plenum Press, New York, 1970. Reproduced with permission.]

synthesis. When such cultures, initially "starved" of serum, were induced to synthesize DNA by the addition of serum, deoxyribonucleoside triphosphate pools expanded three- to fivefold, in coincidence with an increasing rate of DNA synthesis; as the latter declined, so did these pools become smaller. The dCTP pool size was about 30 times that of dGTP in these cells.

Erythrocytes have been a favorite material for biochemical study and considerable data on nucleotide composition are available (see review 22). The nucleotide composition of human erythrocytes is apparent in the chromatographic analysis shown in Fig. 1-5 (39). It is seen that the adenosine phosphates are the most abundant nucleotide species and the pyrimidine ribonucleotides are noteworthy for their virtual absence. Human erythrocytes average about 0.83 mμmole of ATP per milliliter of cells (39). Bartlett has summarized the ATP content of erythrocytes of various species (39).

References

1. Beaven, G. H., Holiday, E. R., and Johnson, E. A., in "The Nucleic Acids" (E. Chargaff and J. N. Davidson, eds.), Vol. 1, p. 493. Academic Press, New York, 1955.
2. Caldwell, I. C., J. Chromatogr. 44, 331 (1969).
3. Jones, W., "Nucleic Acids—Their Chemical Properties and Physiological Conduct," 2nd ed. Longmans, Green, New York, 1970.
4. Levene, P. A., and Bass, L. W., "Nucleic Acids." Chem. Catalog Co. (Tudor), New York, 1931.
5. Michelson, A. M., "The Chemistry of Nucleosides and Nucleotides." Academic Press, New York, 1963.
6. Hurlbert, R. B., Schmitz, H., Brumm, A. F., and Potter, V. R., J. Biol. Chem. 209, 23 (1954).
7. Schmitz, H., Hurlbert, R. B., and Potter, V. R., J. Biol. Chem. 209, 41 (1954).
8. Schmitz, H., Potter, V. R., Hurlbert, R. B., and White, D. M., Cancer Res. 14, 66 (1954).
9. Skoog, L., and Nordenskjöld, B., Eur. J. Biochem. 19, 81 (1971).
10. Potter, R. L., Schlessinger, S., Buettner-Janusch, V., and Thompson, L., J. Biol. Chem. 226, 381 (1957).
11. LePage, G. A., J. Biol. Chem. 226, 135 (1957).
12. Hoff-Jorgensen, E., in "Methods in Enzymology" (S. P. Colowick and N. O. Kaplan, eds.), Vol. 3, p. 781. Academic Press, New York, 1957.
13. Larsson, A., J. Biol. Chem. 238, 3414 (1963).
14. Colby, C., and Edlin, G., Biochemistry 9, 917 (1970).
15. Skoog, L., Eur. J. Biochem. 17, 202 (1970).
16. Lindberg, U., and Skoog, L., Anal. Biochem. 34, 152 (1970).
17. Henderson, J. F., and LePage, G. A., Chem. Rev. 58, 645 (1958).
18. Bock, R. M., in "The Enzymes" (P. D. Boyer, H. Lardy, and K. Myrbäck, eds.), Vol. 2, p. 3. Academic Press, New York, 1960.

19. Leloir, L. F., and Cardini, C. E., *in* "The Enzymes" (P. D. Boyer, H. Lardy, and K. Myrbäck, eds.), 2nd rev. ed., Vol. 2, Part A, p. 39. Academic Press, New York, 1960.

20. Kennedy, E. P., *in* "The Enzymes" (P. D. Boyer, H. Lardy, and K. Myrbäck, eds.), 2nd rev. ed., Vol. 2, Part A, p. 63. Academic Press, New York, 1960.

21. Utter, M. F., *in* "The Enzymes" (P. D. Boyer, H. Lardy, and K. Myrbäck, eds.), 2nd rev. ed., Vol. 2, Part A, p. 75. Academic Press, New York, 1960.

22. Mandel, P., *Progr. Nucl. Acid Res. Mol. Biol.* **3**, 299 (1964).

23. Neufeld, E. F., and Hassid, W. Z., *Advan. Carbohyd. Chem.* **18**, 309 (1963).

24. Neufeld, E. F., and Ginsburg, V., *Annu. Rev. Biochem.* **34**, 297 (1965).

25. Lieberman, I., *J. Amer. Chem. Soc.* **77**, 3373 (1955).

26. Beavo, J. A., Hardman, J. G., and Sutherland, E. W., *J. Biol. Chem.* **245**, 5649 (1970).

27. Carlsson, D. M., and Hansen, R. G., *J. Biol. Chem.* **237**, 1260 (1962).

28. Fox, J. J., Watanabe, K. A., and Bloch, A., *Progr. Nucl. Acid Res. Mol. Biol.* **5**, 251 (1966).

29. Cohen, S. S., *Progr. Nucl. Acid Res. Mol. Biol.* **5**, 2 (1966).

30. Suhadolnik, R. J., "Nucleoside Antibiotics." Wiley, New York, 1970.

31. Nazar, R. N., Lawford, H. G., and Wong, J. T.-F., *Anal. Biochem.* **35**, 305 (1970).

32. Chance, B., Schoener, B., Krejci, K., Russmann, W., Wesemann, W., Schnitger, H., and Bücher, T., *Biochem. Z.* **341**, 325 (1965).

33. Bergmeyer, H. U., "Methoden der enzymatichen Analysen." Verlag Chemie, Weinheim, 1962.

34. Grav, H. J., *Methods Cancer Res.* **3**, 243 (1967).

35. Cohn, W. E., *in* "Methods in Enzymology" (S. P. Colowick and N. O. Kaplan, eds.), Vol. 3, p. 724. Academic Press, New York, 1957.

36. Bucher, N. L. R., and Swaffield, M. N., *Biochim. Biophys. Acta* **129**, 445 (1966).

37. Rotherham, J., and Schneider, W. C., *J. Biol. Chem.* **232**, 853 (1957).

38. Gross, N., and Rabinowitz, M., *Biochim. Biophys. Acta* **157**, 648 (1968).

39. Bartlett, G.. *Advan. Exp. Biol. Med.* **6**, 245 (1970).

CHAPTER 2

CONFIGURATION AND CONFORMATION OF NUCLEOSIDES AND NUCLEOTIDES

I. Introduction

There is a tendency to visualize nucleoside and nucleotide structures as they are written on paper, and not as they exist in the aqueous environment of their biological occurrence and function. This chapter therefore reviews very briefly current knowledge of the configuration and conformations of nucleosides and nucleotides, with an emphasis, wherever possible, on their steric form in aqueous solution. The literature through 1968 is listed and discussed in the review of Preobrazhenskaya and Shabarova (1); see also Jordan (2).

II. The Sugar

A. FURANOSE RING CONFORMATION

The majority of naturally occurring nucleosides and nucleotides are
N-glycosides of β-D-ribofuranose or 2-deoxy-β-D-ribofuranose.

The furanose ring is not planar, but may exist in stable conformations
in which either the C-2 or the C-3 occupy positions which are 0.5 to 0.6
Å away from the major plane of the ring; this displacement may be either
up toward C-5, (*endo*), or down away from C-5 (*exo*). These conformations are depicted in Fig. 2-1. Nucleosides most frequently assume
endo conformations; in solution, purine nucleosides show a preference
for the 2′-*endo* form, whereas pyrimidine nucleosides prefer the 3′-*endo*
form. In crystals both forms coexist, although deoxyadenosine appears
to favor the 3′-*exo* conformation (*3*).

B. *cis*-DIOL SYSTEM OF RIBOSE

The length of the C-2′—C-3′ bond in nucleosides is very close to that
found in ordinary C—C bonds, and the van der Waal's radii of the hydroxyl oxygens at these positions are too large for them to exist in the
eclipsed conformation. The dihedral angle ϕ_{OO} between the C-2′—O-2′
and the C-3′—O-3′ is therefore not 0° as in true *cis* structures, but is 43
to 54°. This is illustrated in Fig. 2-2.

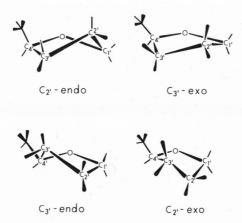

FIG. 2-1. *Endo* and *exo* forms of the furanose ring.

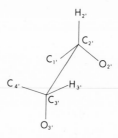

FIG. 2-2. Conformation of the *cis*-glycol groups of ribose.

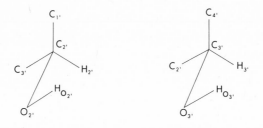

FIG. 2-3. Conformation of the C-2′ and C-3′ hydroxyl groups.

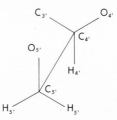

FIG. 2-4. Conformations around the C-4′—C-5′ bond.

FIG. 2-5. Conformation of the 5′-phosphate.

The preferred conformation of the O—H bond of the 2'-hydroxyl group is *trans* to the C-2'—C-3' bond and *gauche* (i.e., at 60°) to the C-1'—C-2' bond. That of the 3'-hydroxyl group is *trans* to the C-2'—C-3' bond and *gauche* to the C-3'—C-4' bond (see Fig. 2-3).

C. The C-4'—C-5' Bond and 5'-Phosphate

The most common conformation of the C-5'—O-5' bond appears to be that which is *gauche* relative both to the C-3'—C-4' bond and to the C-4'—O-4' bond, as shown in Fig. 2-4.

In 5'-nucleotides, the O-5'—P bond is *trans* to the C-4'—C-5' bond, thus assuming the most extended form of the compound (Fig. 2-5).

III. The Bases

Both purine and pyrimidine ring systems are essentially planar, although some small deviations from planarity have been detected, for example, in cytidine.

Some substituents on the heterocyclic rings do not lie in the average plane of the ring, however. The C-1' atoms of the ribofuranose of nucleosides are displaced 0.2 to 0.3 Å, for example.

At physiological pH values, adenine, guanine, cytidine, and their derivatives occur predominantly in the amino tautomeric form, and hypoxanthine, guanine, cytidine, and thymine and their derivatives are predominantly in the keto tautomeric form.

In monoprotonated purine and pyrimidine derivatives, the proton is on N-1 in adenine, N-7 in guanine, and N-3 in pyrimidines (*2, 4*).

IV. Base–Sugar Relationships

The planes of all nucleoside bases are essentially perpendicular to the planes of the nucleoside sugars. However, the bases can rotate about the glycosidic bond, and the relative positions of bases and sugars in this regard are defined by the so-called torsion angle, ϕ_{CN} (*5*). This is the dihedral angle between the plane of the base and the plane formed by C-1', O-1', and N-9 in purines, and C-1', O-1', and N-1 in pyrimidines (see Fig. 2-6). This angle is zero when C-1'—O-1' and N-1—C-2 (in pyrimidines) or N-9—C-4 (in purines) are *trans* or antiplanar. When the base rotates clockwise as one looks from C-1' to N-9 or N-1, the dihedral angle is negative, and when it rotates counterclockwise this angle is positive.

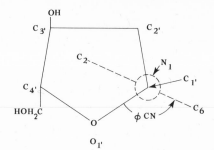

FIG. 2-6. The torsion angle in a pyrimidine deoxyribonucleoside. The observer is looking from C-1' to N-1.

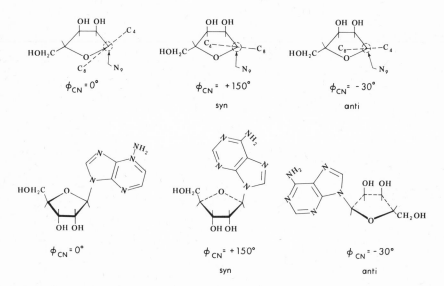

FIG. 2-7. *Syn* and *anti* conformations of adenosine. Light lines denote bonds in the plane of the paper; dotted lines, those behind this plane; and heavy lines, those in front of this plane.

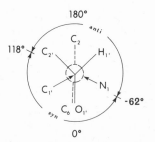

FIG. 2-8. Proposed ranges of ϕ_{CN} for *syn* and *anti* conformation of a pyrimidine nucleoside.

Relatively stable conformations may exist when ϕ_{CN} is about $-30°$, and when it is about $+150°$; these have been defined as the *anti* and *syn* forms, respectively (Fig. 2-7). Recently (*6*), the suggestion has been made that torsion angles between $-62°$ and $+118°$ should be designated *anti*, and all others, *syn* (Fig. 2-8). The barrier between *syn* and *anti* conformations is determined by steric interactions between the N-3 of purine bases or the O-2 of pyrimidine bases, and the oxygen of the pentose ring. In almost every case studied, the *anti* conformation (*1*, *7–11*) is preferred in solution, although both forms are found in crystals (e.g., *3*).

References

1. Preobrazhenskaya, N. N., and Shabarova, Z. A., *Russ. Chem. Rev.* **38**, 111 (1969).
2. Jordan, D. O., "The Chemistry of Nucleic Acids," pp. 97–111. Butterworth, London, 1960.
3. Haschemeyer, A. E. V., and Rich, A., *J. Mol. Biol.* **27**, 369 (1967).
4. Voet, D., and Rich, A., *Progr. Nucl. Acid Res. Mol. Biol.* **10**, 183 (1970).
5. Donohue, J., and Trueblood, K. N., *J. Mol. Biol.* **2**, 363 (1960).
6. Kang, S., *J. Mol. Biol.* **58**, 297 (1971).
7. Schweizer, M. P., Broom, A. D., Ts'o, P. O. P., and Hollis, D. P., *J. Amer. Chem. Soc.* **90**, 1042 (1968).
8. Chan, S. I., and Nelson, J. N., *J. Amer. Chem. Soc.* **91**, 168 (1969).
9. Smith, I. C. P., Blackburn, B. J., and Yamane, T., *Can. J. Chem.* **47**, 513 (1969).
10. Rogers, G. T., and Ulbricht, T. L. V., *Biochem. Biophys. Res. Commun.* **39**, 414 (1970).
11. Rogers, G. T., and Ulbricht, T. L. V., *Biochem. Biophys. Res. Commun.* **39**, 419 (1970).

CHAPTER 3

FUNCTIONS OF NUCLEOTIDES

I. Introduction

The large number of purine and pyrimidine nucleotides found in nature suggests that these compounds will also have a large number of functions in cells, and might participate in many areas of metabolism. Similarly, the quantitative predominance of ATP among all of these compounds would seem to suggest a special importance for this nucleotide in intermediary metabolism. Both conclusions appear to be justified, and they will be illustrated in more detail in this chapter. Because a detailed consideration of nucleotide function would cover almost the whole field of biochemistry, this survey will be brief and will stress *types* of reactions, and will generalize broadly.

Five general types of functions of nucleotides can be distinguished. The first is their role in the storage and "carrying" of metabolically available energy and in the support of energetically unfavorable reactions. This role is mainly that of the adenine nucleotides. In the second type of function the nucleotide acts as carrier of a wide variety of groups and is their donor to appropriate acceptors. The third type of function is that in which nucleotides form structural units in macromolecules or form integral components of low molecular weight compounds. "Physiological" functions, a fourth type, are those in which the effect of the nucleotide is observed at the whole cell, organ, or organism level. Finally, the role of nucleotides in the regulation of intermediary metabolism may be considered a fifth type of function (general references: *1–7*).

II. Role of Nucleotides in Energy Metabolism

A. SYNTHESIS OF HIGH-ENERGY NUCLEOTIDES

The free energy derived from biological oxidations is largely converted into chemical energy in the form of ATP, which is generally considered to be the most important mobile carrier of metabolically available energy in cells.

1. Phosphorylation Linked to Electron Transport

At present, neither oxidative phosphorylation nor photophosphorylation is completely understood. In oxidative phosphorylation, electrons are derived from substrates that enter the Krebs cycle and from fatty acids, whereas in photophosphorylation they are furnished by chlorophyll in the presence of light. ATP synthesis is coupled to the subsequent suc-

TABLE 3–I

TYPES OF SUBSTRATE-LEVEL OXIDATIVE PHOSPHORYLATION REACTIONS

(a) Xylulose-5-P + P_i → acetyl-P + glyceraldehyde-3-P
Acetyl-P + ADP → ATP + acetate

     H acceptor: C-1 of substrate
     Acyl acceptor: P_i
     Enzymes: 2

(b) Acetaldehyde + NAD^+ + CoA → acetyl-CoA + NADH
     Acetyl-CoA + P_i → acetyl-P + CoA
     Acetyl-P + ADP → ATP + acetate

     H acceptor: NAD^+
     Acyl acceptors: CoA, P_i
     Enzymes: 3

(c) Glyceraldehyede-3-P + NAD^+ → 3-P-glyceryl-S-enzyme + NADH
+ enzyme-SH
     3-P-Glyceryl-S-enzyme + P_i → 3-P-glyceryl-P + enzyme-SH
     3-P-Glyceryl-P + ADP → ATP + 3-P-glycerate

     H acceptor: NAD^+
     Acyl acceptors: enzyme-SH, P_i
     Enzymes: 2

(d) Pyruvate + enzyme-thiamine-PP → CO_2 + enzyme-hydroxyethyl-
                                 thiamine-PP
     Flavin + enzyme-hydroxyethyl- → acetyl-P + enzyme-thiamine-
     thiamine-PP + P_i PP + reduced flavin
         Acetyl-P + ADP → ATP + acetate

     H acceptors: oxygen on C-2, flavin
     Aldehyde acceptor: enzyme-thiamine-PP
     Acyl acceptor: P_i
     Enzymes: 2(?)

(e) Pyruvate + enzyme-thiamine-PP → CO_2 + enzyme-hydroxyethyl-
                                 thiamine-PP
     Enzyme-hydroxyethylthiamine- → Acetyl-CoA + enzyme-
     PP + ferridoxin + CoA   thiamine-PP + reduced
                              ferridoxin
         Acetyl-CoA + P_i → acetyl-P + CoA
         Acetyl-P + ADP → acetate + ATP

     H acceptors: oxygen on C-2, ferridoxin
     Aldehyde acceptor: enzyme-thiamine-PP
     Acyl acceptors: CoA, P_i
     Enzymes: 3

TABLE 3–I   (Continued)

(f)   Pyruvate + enzyme-thiamine-PP → CO_2 + enzyme-hydroxyethyl-
thiamine-PP
Enzyme-hydroxyethylthiamine-PP → acetyl-S-enzyme-SH
+ enzyme-S_2 + enzyme-thiamine-PP
Acetyl-S-enzyme-SH + CoA → acetyl-CoA + enzyme-$(SH)_2$
Acetyl-CoA + P_i → acetyl-P + CoA
Acetyl-P + ADP → acetate + ATP

H acceptors: oxygen on C-2, enzyme-S_2
Aldehyde acceptor: enzyme-thiamine-PP
Acyl acceptors: enzyme-S_2, CoA, P_i
Enzymes: 4
Two further reactions are coupled to those
above to regenerate enzyme-S_2:
Enzyme-$(SH)_2$ + FAD → enzyme-S_2 + FADH
FADH + NAD → FAD + NADH

---

cessive oxidation–reduction of members of electron transport chains which pass electron pairs through stages of successively lower potential energy until they reach oxygen, the terminal electron sink.

2. Substrate-Level Oxidative Phosphorylation

In general this process proceeds through the oxidative conversion of a carbonyl group to an acyl derivative which eventually becomes an acyl phosphate. The latter then reacts with ADP to form ATP. Variations exist (a) in the number of oxidations; (b) in the number and type of hydrogen acceptors; (c) in the number and type of acyl derivatives; and (d) in the number of enzymes involved. Examples of several variants are given in Table 3–I; these are discussed at greater length (and in a slightly different classification) by Racker (8).

The substrate-level oxidative phosphorylation system comprised of enolase plus pyruvate kinase differs in several respects from those given in Table 3–I.

2-P-Glycerate → H_2O + P-enol-pyruvate

P-Enol-pyruvate + ADP → ATP + pyruvate

The oxidation of C-2 (for which the hydrogen acceptor is C-3) results in the formation of a high-energy phosphate intermediate. The phosphate has been introduced before the oxidation and comes ultimately from ATP rather than from inorganic phosphate.

B. Transfer of Energy Stored in Nucleotides

Hydrolysis of the pyrophosphate bonds in high-energy nucleoside triphosphates (of which ATP is most important) generates free energy which is made available, directly or indirectly, to drive energetically unfavorable reactions. The energy released in the cleavage of phospho-anhydride linkages is not of concern in the present context; instead we must consider the overall change in free energy when a compound such as ATP transfers one of its substituent groups to another molecule. The free energy change depends in this case on the nature of the acceptor as well as of the donor.

Nucleoside triphosphate-utilizing reactions may be classified in the following groups depending on the site of cleavage of the nucleoside triphosphate molecule and on the nature of the group transferred. In both biological and chemical systems, the α- and γ-phosphorus atoms are more susceptible to nucleophilic attack than is the β-phosphorus.

1. Phosphoryl Group Transfer

Phosphoryl group transfer is catalyzed by kinases (ATP-requiring phosphotransferases), of which a large number are known. The reaction involves nucleophilic attack by the acceptor on the γ-phosphorus of ATP, with consequent cleavage between this phosphorus and the bridge oxygen atom. A divalent cation such as Mg^{2+} is required, and the dominant species present are $MgATP^{2-}$ (substrate) and ADP^{3-} plus acceptor-PO_3^{2-} (products). A number of such reactions require or are activated by K^+.

This group of enzyme reactions may be subdivided according to the nature of the acceptor molecule, as shown in Table 3–II.

2. Pyrophosphoryl Group Transfer

The transfer of the terminal pyrophosphoryl moiety of ATP is known to occur in only two cases. In both reactions Mg^{2+} is required to form $MgATP^{2-}$, and phosphate is required as an activator; its precise role is not yet known.

$$\text{Ribose-5-P} + \text{ATP} \xrightarrow{\text{Mg}^{2+},\ \text{Pi}} \text{5-P-ribose-1-PP} + \text{AMP}$$

$$\text{Thiamine} + \text{ATP} \xrightarrow{\text{Mg}^{2+},\ \text{Pi}} \text{thiamine-PP} + \text{AMP}$$

TABLE 3–II

PHOSPHORYL GROUP TRANSFER REACTIONS (CLASSIFIED ACCORDING
TO TYPES OF ACCEPTOR MOLECULES)

(a) Alcohols

$$\text{Glucose} + \text{ATP} \xrightarrow{\text{Mg}^{2+}} \text{glucose-6-P} + \text{ADP}$$

$$\text{Glycerol} + \text{ATP} \xrightarrow{\text{Mg}^{2+}} \text{glycerol-3-P} + \text{ADP}$$

$$\text{Riboflavin} + \text{ATP} \xrightarrow{\text{Mg}^{2+}} \text{riboflavin-5'-P} + \text{ADP}$$
(Note that only primary alcohols are substrates.)

(b) Hemiacetals (sugars)

$$\text{Galactose} + \text{ATP} \xrightarrow{\text{Mg}^{2+}} \text{galactose-1-P} + \text{ADP}$$

(c) Carboxylic acids

$$\text{Acetate} + \text{ATP} \xrightarrow{\text{Mg}^{2+}} \text{acetyl-P} + \text{ADP}$$

$$\text{Glycerate} + \text{ATP} \xrightarrow{\text{Mg}^{2+}} \text{glyceryl-P} + \text{ADP}$$

(d) Amidines

$$\text{Creatine} + \text{ATP} \xrightarrow{\text{Mg}^{2+}} \text{creatine-P} + \text{ADP}$$

$$\text{Arginine} + \text{ATP} \xrightarrow{\text{Mg}^{2+}} \text{arginine-P} + \text{ADP}$$

(e) Phosphates (see also Chapter 4)

$$\text{GMP} + \text{ATP} \xrightarrow{\text{Mg}^{2+}} \text{GDP} + \text{ADP}$$

$$\text{5-Phosphomevalonate} + \text{ATP} \xrightarrow{\text{Mg}^{2+}} \text{mevalonate-5-PP} + \text{ADP}$$

3. Nucleotidyl Group Transfer

Nucleotide coenzymes involved in group-transfer reactions (see below, Section IV) and similar compounds are frequently formed by reaction of a nucleoside triphosphate with a phosphorylated compound to give pyrophosphate and a product of the type, nucleoside diphospho-X. Pyrophosphate is released from the nucleoside triphosphate following nucleophilic attack on the α-phosphorus by the oxygen of the substrate phosphate.

$$\text{Nicotinamide ribonucleotide} + \text{ATP} \rightarrow \text{Nicotinamide adenine dinucleotide} + \text{PP}_i$$

$$\text{Alcohol-P} + \text{CTP} \rightarrow \text{CDP-alcohol} + \text{PP}_i$$

$$\text{Sugar-1-P} + \text{UTP} \rightarrow \text{UDP-sugar} + \text{PP}_i$$

$$\text{Pantotheine-4-P} + \text{ATP} \rightarrow \text{dephospho-coenzyme A} + \text{PP}_i$$

$$\text{Nicotinic acid ribonucleotide} + \text{ATP} \rightarrow \text{nicotinic acid adenine dinucleotide} + \text{PP}_i$$

The first three nucleotide products are active coenzymes; the latter two are converted to active coenzymes by subsequent reactions (see below, Section III).

The adenyl transfer reaction is also frequently involved in the activation of acyl groups, but in these cases the resulting acyl adenylate is usually enzyme-bound and is broken down by the same enzyme which synthesizes it. In contrast to the nucleotidyl derivatives mentioned above, these derivatives contain only one phosphate, although they are also anhydrides; thus

$$\text{Acetate} + \text{ATP} \rightarrow \text{acetyl-AMP} + \text{PP}_i$$

$$\text{Lysine} + \text{ATP} \rightarrow \text{lysyl-AMP} + \text{PP}_i$$

A very few stable nucleoside monophosphate coenzymes are formed through nucleotidyl transfer reactions.

$$\text{SO}_4^{2-} + \text{ATP} \rightarrow \text{AMP-SO}_4^- + \text{PP}_i$$

$$\text{Neuraminic acid} + \text{CTP} \rightarrow \text{CMP-neuraminic acid} + \text{PP}_i$$

Finally, because pyrophosphate is a product, the adenyl cyclase reaction may be included here.

$$\text{ATP} \rightarrow 3',5'\text{-cyclic AMP} + \text{PP}_i$$

The anhydride bonds formed in nucleotidyl transfer reactions are usually equal to or greater in free energy of hydrolysis than the anhydride bond of ATP which was broken; however, these reactions are driven to completion by the cleavage of pyrophosphate by pyrophosphatase. The latter is a reaction of high free energy, and this makes the synthetic process virtually irreversible.

4. 5'-Deoxyadenosyl Group Transfer

This reaction involves cleavage of the carbon–oxygen bond at the 5'-position of ATP. In the synthesis of 5'-deoxyadenosylcobalamin (or coenzyme B_{12}), the deoxynucleoside moiety replaces the cyanide of cyanocobalamin, and trimetaphosphate is formed. Details of this reaction are still unclear.

S-Adenosylmethionine is formed in a similar reaction.

$$\text{ATP} + \text{methionine} \rightarrow S\text{-adenosylmethionine} + PP_i + P_i$$

Trimetaphosphate is formed, but before being released from the enzyme surface, it is cleaved to phosphate (the γ-phosphate of ATP) and pyrophosphate (the α- and β-phosphates of ATP).

C. Utilization of Energy Stored in Nucleotides

The reaction types outlined above indicate the various ways in which the free energy of ATP can be released through hydrolysis of one or more of its phosphoanhydride bonds. This energy can be used in several different ways to drive endergonic reactions.

1. Synthesis of Other Nucleoside Triphosphates

ATP is the primary source of utilizable energy for biosynthetic reactions, but other purine and pyrimidine nucleoside triphosphates (GTP, UTP, CTP) occur in cells and are formed by phosphoryl transfer from ATP, as described in detail in Chapter 4. These are also high-energy compounds, and their hydrolysis can be coupled to energetically unfavorable reactions. This most commonly takes place through nucleotidyl transfer reactions leading to group-transfer coenzymes (Section II, B, 3).

2. Synthesis of Other High-Energy Substrates

The phosphoanhydride bond energy of ATP can also participate indirectly in endergonic reactions through the synthesis of substrates other than nucleoside triphosphates whose cleavage releases sufficient energy to drive the reaction. For example, the products of almost all of the nucleotidyl and deoxynucleosidyl transfer reactions described above are high-energy compounds, as are the products of some of the phosphoryl and pyrophosphoryl transfer reactions. In all of these reactions, the group originally donated by ATP or another nucleoside triphosphate is released in the course of the endergonic reaction and does not appear in the product.

In those cases in which the products of phosphoryl transfer reactions (e.g., glucose-6-P, ribose-5-P) are not high-energy compounds, high or moderately high energy compounds (e.g., glucose-1-P, ribose-1-P) may be formed by one or more further transformations which do not require ATP.

3. Direct Use of Adenosine Triphosphate

A large group of biosynthetic enzyme reactions, now called "ligases," use ATP as an immediate substrate, but no part of the ATP appears attached to any of the products. In terms of the way in which ATP is used, these reactions may be divided into two subclasses.

(a) $ATP \rightarrow AMP + PP_i$

In this type of reaction an intermediate enzyme-bound acetyl (acyl) adenylate is believed to be formed; this subsequently reacts with some acceptor of the acetyl group, such as CoA:

$$ATP + acetate + CoA \rightarrow AMP + PP_i + acetyl\text{-}CoA$$

(b) $ATP \rightarrow ADP + P_i$

In the second type of reaction the existence of either a phosphorylated enzyme intermediate or of an enzyme-bound, phosphorylated-substrate intermediate is inferred. A phosphorylated enzyme has been detected in the succinyl-CoA synthetase reaction, for example,

$$ATP + succinate + CoA \rightarrow ADP + P_i + succinyl\text{-}CoA$$

In another case, γ-glutamyl phosphate is believed to be an enzyme-bound intermediate in the glutamine synthetase reaction.

$$ATP + NH_3 + glutamate \rightarrow ADP + P_i + glutamine$$

4. Direct Use of Guanosine Triphosphate

In a few reactions, GTP serves as nucleoside triphosphate substrate:

$$Oxaloacetate + GTP \rightarrow phosphoenolpyruvate + CO_2 + GDP$$

$$IMP + aspartate + GTP \rightarrow adenylosuccinate + GDP + P_i$$

GTP is also essential for polypeptide synthesis in cell-free systems. The mechanism for the GTP utilization in protein synthesis is not yet clear, although GTP may supply the energy needed for the coordinated reactions involving the binding of amino acyl-tRNA, formation of the peptide bond and the movement of messenger RNA along ribosomes.

III. Transformation of Nucleotide Coenzymes

In some cases a nucleotidyl derivative may be further modified either
to form an active group-transfer coenzyme, or to form a different active
coenzyme. This applies particularly to the nucleotide sugars, where very
significant transformations occur. The types of reaction which occur are
given in Table 3–III.

IV. Group-Transfer Reactions of Nucleotide Coenzymes

Section II,B of this chapter considered reactions in which parts of
nucleoside triphosphates were transferred to some acceptor, whereas this
section considers the transfer of other groups attached to nucleoside mono-
or diphosphate carriers, to appropriate acceptors, most of which are not
nucleotides. The number of compounds transferred in this way is very
large and only general types of groups, and types of products formed, will
be outlined here.

A. GLYCOSYL GROUP TRANSFER

Monosaccharides must be activated in order to participate in oligosac-
charide synthesis, to lengthen the chains of polysaccharides, or combine
with hydroxyl, amino, or carboxyl groups of non-sugar acceptors. Nucleoside diphosphate sugars are the preferred substrates for such reactions and
have higher free energies of hydrolysis than other types of glycosyl compounds.

Uracil is the predominant base in the nucleoside diphosphate sugars of
higher plants and animals, although sugar nucleotides with other bases
also occur in these organisms. In microorganisms and lower plants, however,
sugar nucleotides with other bases are more widely used. Many glycosyl
transfer enzymes are not specific with respect to the base used, and it is
sometimes difficult to establish exactly which sugar nucleotide is of primary
physiological importance. The reviews of Hassid (*10,11*) should be consulted.

1. Oligosaccharide Synthesis

Sucrose:

$$\text{UDP-glucose} + \text{fructose} \leftrightarrow \text{sucrose} + \text{UDP}$$

$$\text{UDP-glucose} + \text{fructose-6-P} \rightarrow \text{sucrose-P} + \text{UDP}$$

Phosphate is subsequently removed by hydrolysis of the product of the

TABLE 3–III

TYPES OF NUCLEOTIDE COENZYME TRANSFORMATION REACTIONS

(a) Phosphorylation
Nicotinamide adenine dinucleotide + ATP → nicotinamide adenine dinucleotide
2'-phosphate + ADP
Dephospho-coenzyme A + ATP → coenzyme A + ADP
Adenosine phosphosulfate + ATP → 3'-phosphoadenosine phospho-
sulfate + ADP

(b) Adenylation of an already active coenzyme
Riboflavin-5-P + ATP → flavin adenine dinucleotide + PP_i

(c) Amidation
Nicotinic acid adenine dinucleotide → nicotinamide adenine dinucleotide
+ glutamine + ATP   + glutamate + P_i + ADP

(d) Epimerization
(i) At the 4-position of hexoses
UDP-glucose + NAD^+ → UDP-galactose + NAD^+
(ii) At the 2-position
$$\text{UDP-}N\text{-acetylglucosamine} \xrightarrow{\ NAD^+\ } \text{UDP-}N\text{-acetylmannosamine}$$
(iii) At the 5-position
$$\text{UDP-glucuronate} \xrightarrow{\ NAD^+\ } \text{UDP-iduronate}$$

(e) Oxidation of sugars to uronic acids
UDP-glucose + 2 H_2O + 2 NAD^+ → UDP-glucuronate + 2 H^+ + 2 NADH

(f) Reduction of sugars to deoxy sugars
(i) Monodeoxy sugars
GDP-mannose → GDP-4-ketofucose → GDP-4-keto-6-deoxymannose →
GDP-fucose
(ii) Dideoxy sugars
CDP-glucose → CDP-4-keto-deoxyglucose → CDP-3,6-dideoxyglucose

(g) Decarboxylation of uronic acids
UDP-glucuronate → UDP-xylose + CO_2

(h) Synthesis of branched-chain sugars
TDP-4-ketorhamnose → TDP-streptose

(i) Sulfate removal (9)
Phosphoadenosine phosphosulfate → phosphoadenosine phosphate + enzyme-
bound sulfate
Adenosine phosphosulfate → AMP + enzyme-bound-flavin-sulfate

second reaction. It has been suggested that the first reaction provides for the degradation of sucrose, whereas sucrose is synthesized by the second.

Trehalose, a disaccharide of glucose, is synthesized by yeast and certain insects in a similar fashion:

$$\text{UDP-glucose} + \text{glucose-6-P} \rightarrow \text{trehalose-P} + \text{UDP}$$

Lactose (in milk and lactating mammary tissue):

$$\text{UDP-galactose} + \text{glucose} \rightarrow \text{lactose} + \text{UDP}$$

Raffinose (in plant seeds):

$$\text{UDP-galactose} + \text{sucrose} \rightarrow \text{raffinose} + \text{UDP}$$

2. Glycoside Synthesis

The functions of the many and varied plant glycosides are unclear; they do not serve as major food reserves or as structural material. Animal cells, particularly in liver and kidney, convert numerous alcohols, amines, and carboxylic acids to β-glucuronides, which can be transported in the blood and excreted in urine or bile. This is a major route of detoxication of foreign substances. In plants, glucuronides are rare, and glucosides are the predominant sugar derivatives. (See reference 12.)

$$\text{UDP-glucose} + \text{hydroquinone} \rightarrow \text{UDP} + \text{hydroquinone glucoside}$$

$$\text{UDP-glucose} + \text{phenol } \alpha\text{-glucoside} \rightarrow \text{UDP} + \text{phenol } \beta\text{-gentiobioside}$$

$$\text{UDP-glucose} + \text{anthranilic acid} \rightarrow o\text{-aminobenzoyl glucoside} + \text{UDP}$$

UDP-glucose + hydroxymethylcytosine → UDP + mono- or diglucosyl
(in DNA of T-even phage)   hydroxymethylcytosine

UDP-glucuronic acid can be transferred to cortisol to form an ether O-glucuronide; bilirubin to form an ester O-glucuronide; aniline to form an amide N-glucuronide; mercaptobenzothiazole to form a thioether S-glucuronide.

3. Homopolysaccharide Synthesis

In the synthesis of homopolysaccharides the monosaccharide units are activated by the formation of nucleoside diphosphate derivatives from which they are transferred to a nonactivated growing polymer. The following nucleotide coenzymes are used in the synthesis of the most common polymers.

Glycogen:

UDP-glucose (most animal cells)
ADP-glucose (many microorganisms)

Starch:

> UDP-glucose
> ADP-glucose

(Either one or the other, or both together, are used in different plants.)

Cellulose:

> GDP-glucose (plants)
> UDP-glucose (*Acetobacter xylinium*)

Chitin:

> UDP-*N*-acetylglucosamine

4. Heteropolysaccharide Synthesis

The synthesis of heteropolysaccharides is somewhat more complex than that of homopolysaccharides, because at each step a different primer is required and a different activated monosaccharide is added.

Hyaluronic acid:

> UDP-*N*-acetylglucosamine and UDP-glucuronic acid
> (alternatively)

Chondroitin:

> UDP-*N*-acetylgalactosamine and UDP-glucuronic acid
> (later sulfated at the polymer level)

Alginic acid:

> GDP-mannuronic acid and GDP-guluronic acid

5. Glycoproteins and Related Compounds

The carbohydrate portion of glycoproteins is formed from UDP-*N*-acetylglucosamine and CMP-*N*-acetylneuraminic acid, and is attached to the protein by *N*-glycosidic bonds to asparagine amide groups.

Mucopolysaccharides, in contrast, are believed to be attached to peptides through serine or threonine hydroxyl groups.

Bacterial cell walls contain mureins which are polymers formed from UDP-*N*-acetylglucosamine and UDP-*N*-acetylmuramyl peptides.

Bacterial cell wall lipopolysaccharides of *Salmonella* are some of the most interesting examples of the coordinated use of sugar nucleotides involving both different sugars and different nucleotide bases. To a backbone of hexose, phosphate, and ketodeoxyoctanoate, are added in a sequential manner the sugars galactose, abequose, rhamnose, mannose, and *N*-acetylglucosamine. These are formed and activated by the following scheme (*13*):

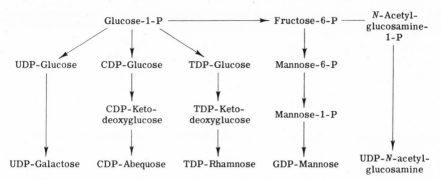

B. Alcohol Phosphate Transfer

Cytidine nucleotides apparently are the only nucleotides involved in the activation and transfer of alcohols. Like many other group-transfer coenzymes, they are formed by an ordinary nucleotidyl transfer reaction:

$$CTP + alcohol\text{-}P \to CDP\text{-}alcohol + PP_i$$

In contrast to many similar reactions, however, alcohol-P is the moiety transferred to the acceptor, rather than just the alcohol itself.

1. Phospholipid Synthesis

There are several ways in which CDP-alcohol coenzymes may be involved in phospholipid synthesis.

$$CDP\text{-}ethanolamine + \alpha,\beta\text{-}diglyceride \to CMP + phosphatidylethanolamine$$

$$CDP\text{-}choline + \alpha,\beta\text{-}diglyceride \to CMP + lecithin$$

$$CDP\text{-}choline + N\text{-}acylsphingosine \to sphingomyelin + CMP$$

$$CDP\text{-}\alpha,\beta\text{-}diglyceride + serine \to phosphatidylserine + CMP$$

2. Teichoic Acid Synthesis

Teichoic acids consist of polyolphosphate homopolymers to which sugars and amino acids are attached; they are found in bacterial cell walls. There are two common types, in which glycerol phosphate and ribitol phosphate, respectively, constitute the repeating units. These are transferred to the appropriate primer through the corresponding nucleotide derivative, CDP-glycerol or CDP-ribitol.

C. Sulfate Transfer

3'-Phosphoadenosine 5'-phosphosulfate is the coenzyme through which sulfate is activated and transferred. Acceptors include sugars polymerized in chondroitin, for example, and alcohol and amino groups in drugs and other foreign substances.

D. HYDROGEN, HYDRIDE ION, AND ELECTRON TRANSFER

1. Pyridine Nucleotide Coenzymes

The reactions of the pyridine nucleotides, NAD^+ and $NADP^+$ are so common and well known that detailed discussion is unnecessary. They probably accept and transfer hydride ions (i.e., proton plus two electrons).

2. Flavin Nucleotide Coenzymes

The flavin nucleotide FAD, and its closely related coenzyme, FMN, carry out one- and two-electron transfers.

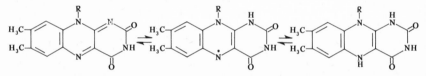

R = (ribitol 5'-phosphate) or (ribitol 5'-phosphate)-(5'''-phosphoadenosine)

3. Coenzyme B_{12}

Coenzyme B_{12} carries out *inter*molecular hydrogen transfer in the ribonucleotide reductase reaction

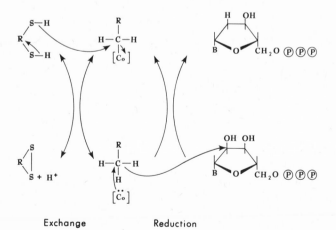

Exchange   Reduction

and *intra*molecular hydrogen transfer in the glycol dehydratase reaction.

E. ALKYL GROUP TRANSFER

1. *S-Adenosylmethionine*

S-Adenosylmethionine (SAM) is the main agent for the transfer of methyl groups, although in some cells choline, betaine, and related compounds also play such a role. Some methyltransferases are very specific with respect to the acceptor substrate while others are unspecific. In all, a very wide variety of acceptors have been identified. The nucleotide product of methylation is *S*-adenosylhomocysteine (SAH). (See reference *14*.) Types and reactions include:

N-Methylation:

> Norepinephrine + SAM → SAH + epinephrine

O-Methylation:

> Epinephrine + SAM → SAH + 3-methoxyepinephrine

S-Methylation:

> 2-Thiouracil + SAM → SAH + 2-methylthiouracil

C-Methylation:

> Unsaturated fatty acids + SAM → SAH + cyclopropane fatty acids

2. *Coenzyme B₁₂*

Coenzyme B_{12} is involved in *intra*molecular alkyl group transfer in the methylmalonyl-CoA isomerase and glutamate mutase reactions .(see reference *14*), both of which may be schematically represented as:

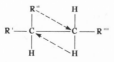

The actual mechanism is uncertain. The methylation of homocysteine and certain other coenzyme B_{12}-requiring reactions appear to involve *inter*molecular alkyl transfer, but the details of these reactions are still obscure.

3. Coenzyme A

Those reactions of coenzyme A in which the α-methylene group of acetyl-CoA is added to an acceptor molecule may also be considered a type of alkyl group transfer. Although this position is activated by being in a CoA ester, the ester group itself is not primarily involved in the new carbon–carbon bond formation.

$$\text{Acetyl-CoA} + \text{oxaloacetate} \rightarrow \text{citrate} + \text{CoA}$$
$$2(\text{Acetyl-CoA}) \rightarrow \text{acetoacetyl-CoA} + \text{CoA}$$

F. Acyl Group Transfer

1. Enzyme-Bound Acyl Adenylates

Enzyme-bound acyl adenylates are common intermediates in the synthesis of other acyl group transfer coenzymes such as acetyl-CoA. In addition, the activation of fatty acids for glyceride synthesis involves acyl adenylates

$$\text{RCOOH} + \text{ATP} \rightarrow \text{RCO-AMP (enzyme-bound)} + \text{PP}_i$$

as does the activation of amino acids for protein synthesis.

$$\underset{\underset{NH_2}{|}}{R-C-COOH} + ATP \longrightarrow \underset{\underset{NH_2}{|}}{R-C-CO\text{-AMP (enzyme-bound)}} + PP_i$$

2. Coenzyme A

Coenzyme A is the coenzyme carrier of non-enzyme-bound acyl groups and participates in a wide variety of reactions. Among the types of acceptor molecules which are acylated with displacement of CoA are arylamines, choline, imidazole, orthophosphate, dihydrolipoate, and HCN. α,β-Enoyl-CoA compounds may also react at the double bond with different acceptors without displacement of the CoA. (See reference 15.)

V. Nucleotides as Structural Units

A. Nucleic Acids

The bulk of the cellular nucleotides are in the form of the polymers RNA and DNA. In each case the polymerization process is simply a nucleotidyl transfer reaction of nucleoside triphosphates to form a chain of nucleoside monophosphates in 3′,5′-phosphodiester linkage, with consequent release of pyrophosphate.

A detailed discussion of the nucleic acid polymerases is beyond the scope of this work. However, a few comments will be made concerning the nucleotide substrates.

1. Deoxyribonucleic Acid

The DNA nucleotidyltransferase of Kornberg is the most studied enzyme in this field, and has been used to synthesize biologically active molecules of phage ϕX 174 DNA. There is some question, however, concerning its exact function in intact cells. (See reference 16.)

DNA nucleotidyltransferase appears to have only one binding site for nucleoside triphosphates, for which all four substrates compete. Not all compete equally well, however, as shown by their dissociation constants: dGTP, 12 μM; dATP, 33 μM; dTTP, 81 μM; dCTP, 147 μM (16). These values were determined in the absence of DNA primer and template and may be different in a complete system.

The triphosphate is bound adjacent to the 3′-hydroxyl group of the terminal nucleotide of the DNA primer and oriented so that it can form a base pair with the template. When the correct base pair has been formed, the 3′-hydroxyl of the growing chain makes a nucleophilic attack on the α-phosphorus of the deoxynucleoside triphosphate.

The basis for specificity of DNA nucleotidyltransferase is probably not in the recognition by the enzyme of an incoming triphosphate, but rather in its requirement for the correct base pair. The correct base will bind to the template, whereas an inappropriate base will not and will dissociate from the enzyme.

2. Ribonucleic Acid

RNA nucleotidyltransferase is believed to be responsible for the synthesis of the bulk of cellular RNA. This reaction requires all four nucleoside triphosphates and a DNA template, only one strand of which is transcribed. No primer is needed. RNA-dependent (i.e., RNA template) RNA nucleotidyltransferases are induced in cells after infection with RNA viruses, but this discussion will deal with the DNA-dependent enzymes only. (See references 17, 18).

RNA synthesis appears to start with a purine ribonucleoside triphosphate. The ratio of GTP to ATP in the initial position varies with the DNA template and is governed by polypyrimidine stretches in the transcribed strand of DNA. Thus, the purine nucleotides appear to have roles both in initiation of chain synthesis and as regular substrates for chain propagation. The initial purine nucleotide can also stabilize the DNA-bound enzyme.

In the absence of Mg^{2+} the two purine nucleoside triphosphates compete for a single binding site, one presumably involved with initiation. With Mg^{2+}, they compete for a second site, and the two pyrimidine nucleoside triphosphates also compete for a single site. The Michaelis constants for the nucleotide substrates are in the range 0.05 to 0.3 μM, but vary with the DNA template.

Polynucleotide phosphorylase catalyzes the formation of RNA from nucleoside diphosphates, with phosphate as the second product. This reaction is reversible, and at the normal cellular concentration of inorganic phosphate and nucleoside diphosphates the equilibrium condition favors phosphorolysis of RNA rather than polymerization. Although the addition of primer RNA stimulates RNA formation, it does not specify the nucleotide sequence of the product. Polynucleotide phosphorylase, therefore, appears to be involved in the degradation of cellular RNA rather than in its formation.

The utilization of nucleoside triphosphates for the synthesis of DNA and RNA in the reactions catalyzed by DNA and RNA nucleotidyl transferases has an advantage over the use of nucleoside diphosphates. The cleavage by pyrophosphatase of the pyrophosphate group released in the polymerase-catalyzed reaction is a reaction of high free energy, and the biosynthetic process is thus virtually irreversible. Nucleoside triphosphates, but not nucleoside diphosphates, therefore provide both driving force and irreversibility to the biosynthetic reactions.

B. HISTIDINE

The pathway of histidine biosynthesis, as it is now known, is shown in Fig. 3-1. The N-1 and C-2 of ATP, and the ribose-P which becomes bound to its N-1 position, provide the skeleton of histidine, while the phosphoribosyl aminoimidazole carboxamide remnant is released. As is shown in Chapter 7, phosphoribosyl aminoimidazole carboxamide is an intermediate in the pathway of purine biosynthesis *de novo*, and so histidine synthesis not only provides a branch leading into this pathway, but actually makes a loop.

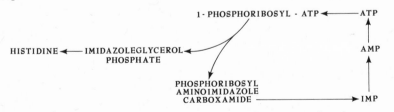

The pathway of histidine biosynthesis is also interesting because it

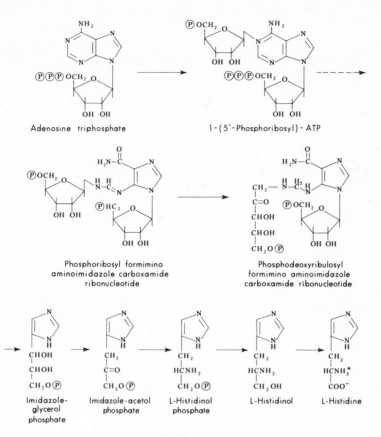

Fig. 3-1. Histidine biosynthesis.

contains phosphoribosyltransferase, glutamine amide transfer, and hetero-cyclic ring closure reactions similar to those in purine biosynthesis itself, as discussed in Chapter 5.

C. Vitamins

Three vitamins are composed in part of heterocyclic rings derived from purine bases or nucleotides. In the plants and microorganisms which make these compounds, therefore, these reactions may be considered either as a nucleotide function or as a new area of nucleotide metabolism.

1. Folic Acid

Figure 3-2 shows that GTP provides the entire 2-amino-4-hydroxy-6-methyldihydropterin moiety of folic acid, although there is still some un-

certainty regarding a number of steps in this pathway. The discovery of
still other intermediates may be anticipated. It may also be pointed out that
several of the reactions involved are similar to those in purine biosynthesis,
including amide synthesis and heterocyclic ring closure (see Chapter 5).

2. Thiamine

The pyrimidine ring moiety of thiamine has recently been shown to be
derived in part from an intermediate in the pathway of purine biosynthesis
de novo, although many details regarding this synthesis still remain unclear.
The proposed scheme is shown in Fig. 3-3. Aspartate may provide the other
part of this ring.

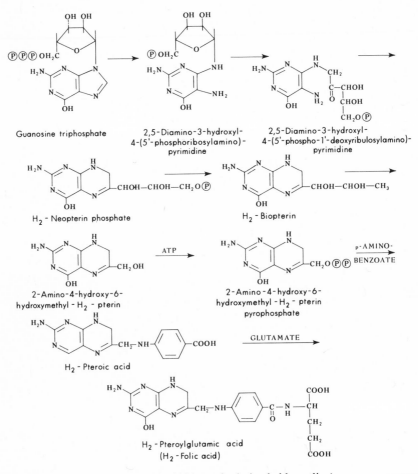

FIG. 3-2. Folic acid biosynthesis (probable outline).

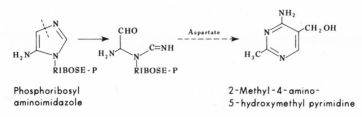

Phosphoribosyl
aminoimidazole

2-Methyl-4-amino-
5-hydroxymethyl pyrimidine

FIG. 3-3. Thiamine pyrimidine synthesis (proposed).

3. Riboflavin

It has been known for some time that addition of purines to incubation media stimulates riboflavin synthesis in certain microorganisms, and studies with labeled purines have shown that, as with folic acid, the C-8 of the purine ring is removed. Whether this takes place at the base, nucleoside, or nucleotide level is unclear. Guanine or xanthine is believed to be a proximal precursor in some cells, whereas hypoxanthine is believed to be involved in others. Figure 3-4 (see p. 50) shows one proposed scheme, although many details remain to be clarified. The biggest remaining question is the identity of the four-carbon unit which adds to the uracil derivative.

VI. Physiological Functions of Nucleotides

A. 3',5'-CYCLIC AMP AND RELATED COMPOUNDS AS MEDIATORS OF HORMONE ACTION

The actions of a number of hormones (Table 3–IV) are mediated in whole or in part, by 3',5'-cyclic AMP, which is regarded as an "intracellular second messenger" (see references 19, 20). Cyclic AMP produces dif-

TABLE 3–IV

Hormones Whose Actions May Be Mediated by
Cyclic AMP[a]

Catecholamines	Triiodothyronine
Glucagon	Melanocyte stimulatory hormone
Acetylcholine	Histamine
Lutenizing hormone	Serotonin
Angiotensin	Insulin
Vasopressin	Prostaglandins
Thyrotropin	

[a] References (19) and (20).

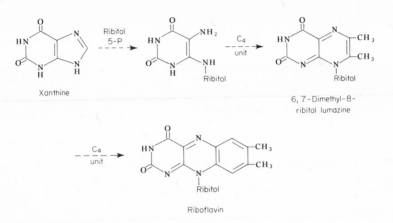

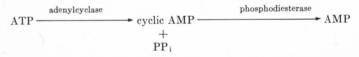

FIG. 3-4. Riboflavin synthesis (proposed early steps).

ferent effects in different cells because the cells themselves are different. The intracellular target systems on which it acts to produce the effects of the primary hormones are not known in all cases, but the effects which cyclic AMP is known at the present time to produce are listed in Table 3–V (see references *19, 20*).

Concentrations of cyclic AMP depend on the activities of two enzymes, adenyl cyclase and a specific phosphodiesterase which catalyzes its hydrolysis to adenylate.

$$\text{ATP} \xrightarrow{\text{adenylcyclase}} \begin{array}{c}\text{cyclic AMP} \\ + \\ \text{PP}_i\end{array} \xrightarrow{\text{phosphodiesterase}} \text{AMP}$$

Adenyl cyclase is membrane-bound, and appears to be the target for hormone action, but the reason for its hormonal sensitivity is not known. Whether the hormonal receptor is actually part of the enzyme, or an independent entity, remains to be established. The mammalian adenyl cyclase is part of, or is closely associated with, the cell membrane or various intracellular membranes.

Recent studies have shown that cyclic AMP-dependent protein kinases are present in many cells, and that these may phosphorylate protamine, histone, casein, phosphorylase kinase, and presumably other proteins as well. Kuo and Greengard (*21*) have proposed that cyclic AMP mediates all of its actions by activating specific protein kinases in various tissues. The tissue-specific effects of cyclic AMP on a given tissue would then be accounted for in terms of the specific protein kinase and its substrates in that tissue. The apparent Michaelis constants for activation by cyclic AMP are from 2×10^{-7} M to 3×10^{-8} M, and it has been suggested that the

TABLE 3-V

CELLULAR PROCESSES INFLUENCED BY CYCLIC AMP[a]

Glycogen phosphorylase	Ketogenesis
Glycogen synthetase	Amino acids → liver protein
Phosphofructokinase	Acetate → liver fatty acids
Fructose 1,6-diphosphatase	Release of amylase
Tyrosine aminotransferase	Release of insulin
Steriodogenesis	Permeability
Lipolysis	Melanocyte dispersion
Glucose oxidation	HCl secretion
Gluconeogenesis	NADPH oscillations
Urea formation	Release of protein
Protein kinase	from polyribosomes

[a] References (19) and (20).

activation is accomplished by means of adenylation of the enzyme by cyclic AMP. Intracellular concentrations of cyclic AMP are in the 10^{-7} to 10^{-6} M range.

The only other 3′,5′-cyclic nucleotide found in nature to date is cyclic GMP. Like cyclic AMP, the guanine derivative is excreted in the urine in relatively large amounts. The amount of cyclic GMP excreted in rat urine does vary with the endocrine status of the animal, but specific hormones affect it differently than cyclic AMP. Cyclic GMP is widely distributed in nature, but is present in tissues in lower concentrations than cyclic AMP. Its synthesis is catalyzed by a cyclase distinct from adenyl cyclase.

The phosphodiesterase which cleaves cyclic AMP to adenylate is found in both particulate and soluble fractions of tissue homogenates, and it is inhibited by pyrophosphate and ATP, as well as by caffeine, theophylline, and theobromine. These three methylated xanthines cause central nervous system and respiratory stimulation, smooth muscle relaxation, diuresis, coronary dilatation, cardiac stimulation, and skeletal muscle stimulation; these effects are believed to be due to increased tissue concentrations of cyclic AMP resulting from inhibition of phosphodiesterase activity.

Caffeine   Theophylline   Theobromine

B. CYTOKININS

Cytokinins are a group of substances which promote cell division and regulate growth in plants in the same manner as does 6-furfurylaminopurine (kinetin) (see references *22*, *23*). A number of cytokinins are naturally occurring adenine derivatives, whereas kinetin itself arises during DNA breakdown under certain conditions; 6-benzylaminopurine is a synthetic cytokinin. The naturally occurring purine cytokinins are listed in Table 3–VI. A closely related compound, triacanthine or 3-(3-methyl-2-butenyl-amino)purine, does not have cytokinin activity.

TABLE 3–VI

NATURALLY OCCURRING CYTOKININS[a]

6-(3-Methyl-2-butenylamino)purine
6-(3-Methyl-2-butenylamino)-9-β-D-ribofuranosylpurine
6-(4-Hydroxy-3-methyl-2-*trans*-butenylamino)purine   (zeatin)
6-(4-Hydroxy-3-methyl-2-*cis*-butenylamino)-9-β-D-
    ribofuranosylpurine
6-(3-Methyl-2-butenylamino)-2-methylthio-9-β-D-
    ribofuranosylpurine
6-(4-Hydroxy-3-methyl-2-butenylamino)-2-methylthio-
    9-β-D-ribofuranosylpurine
6-(3-Methyl-4-hydroxybutylamino)purine

---

[a] See reference (*22*).

The cytokinins are extremely potent; as little as 5×10^{-11} M 6-(4-hydroxy-3-methyl-*trans*-2-butenylamino) purine can be detected by some bioassay systems. They have a wide variety of actions in plants, including promotion of cell division, delay of senescence, and resistance to adverse conditions. They affect plant metabolism, organelle development, and fruit and flower formation. Unfortunately, their basic mechanisms of action are not known.

Zeatin is found in corn and other plants as the free base, 6-(4-hydroxy-3-methyl-*trans*-2-butenylamino)purine, as well as in the form of the ribonucleoside and the ribonucleoside 5'-phosphate. 6-Benzylaminopurine can also be converted to its ribonucleoside and ribonucleoside monophosphate by plant enzymes. Mevalonate is presumed to be a precursor of the natural cytokinins, and cysteine and methionine are thought to be precursors of the thio and methyl portions of the 2-methylthio cytokinin derivatives.

The cytokinins themselves cannot be incorporated into nucleic acids, but adenine residues in tRNA can be converted to cytokinin moieties. Roughly 0.05 to 0.1% of tRNA bases are estimated to be cytokinins. 6-(3-Methyl-2-butenylamino)purine,   6-(4-hydroxy-3-methyl-2-*cis*-butenylamino)purine, and the 2-methylthio derivatives are among those found in tRNA. It must be emphasized that the relation of the cytokinin constituents of tRNA to the plant hormone activity of cytokinins remains to be determined.

C. Effects on Smooth Muscle

1. Adenosine

Adenosine has long been known to cause dilatation of coronary blood vessels, and Berne (24, 25) and others (26) have suggested that adenosine produced within the heart might help to regulate coronary blood flow in this way. Thus, factors which reduce myocardial oxygen tension, such as decreased coronary blood flow, hypoxia, or increased myocardial metabolic activity, accelerate adenine nucleotide breakdown and impair resynthesis. The result is formation of adenosine, which interacts with the vascular smooth muscle cells to cause dilatation and consequently increased blood flow, increased oxygen tension, and removal of adenosine. Although Berne proposed that the adenosine is produced in the myocardial cells and reaches the coronary arterioles via the interstitial space, Baer and Drummond (26) have shown that adenylate can be dephosphorylated rapidly by the coronary vasculature itself.

The duration of coronary dilatation by adenosine is controlled in part by the active adenosine deaminase of the heart, and certain clinically useful vasodilators are believed to act through inhibition of adenosine deamination. Thus dipyridamole (Persantin), or 2,6-bis(diethanolamino)-4,8-dipiperidinopyrimido(5,4-d)pyrimidine is a smooth muscle relaxant used particularly as a coronary vasodilator; it also increases cardiac oxygen consumption. It decreases nucleotide breakdown, causes an accumulation of adenosine in hypoxic heart muscle, and increases blood adenosine concentrations by decreasing its uptake by erythrocytes. Papaverine is also reported to decrease adenosine uptake by erythrocytes, and inhibition of adenosine deamination by ouabain has also been reported.

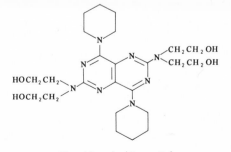

Dipyridamole (Persantin)

2. *Adenine Nucleotides*

Adenine nucleotides also cause smooth muscle to contract, and Daniel and Irwin (*27*) believe that this is due to their effects on calcium ion movement in and out of the muscle. Whether this is a physiologically significant effect is not clear.

D. PLATELET AGGREGATION

In blood platelets approximately 60% of the ADP + ATP is contained in granules together with serotonin (5-hydroxytryptamine). These adenine nucleotides are relatively inert metabolically, as incubation of platelets with radioactive precursors does not lead to their labeling, whereas nonparticulate nucleotides readily become highly labeled (see references *28*, *29*).

One of the first steps in blood clotting is a reversible aggregation of platelets following their interaction with collagen or thrombin. These agents somehow cause the release of the granule-bound ADP + ATP + serotonin from the cells, and it is this ADP which actually causes aggregation. The exact mechanisms both of the release and of the aggregation are not known, although there are numerous hypotheses.

VII. Regulatory Functions of Nucleotides

In addition to the roles of nucleotides and their derivatives and products in intermediary metabolism and in polynucleotide biosynthesis, some nucleotides can also regulate a wide variety of enzyme reactions and metabolic processes; in doing so, they are not themselves metabolized in the reaction. For reviews dealing with the vast literature on metabolic regulation, see references *30–32*.

A. ALLOSTERIC EFFECTS OF ADENINE NUCLEOTIDES

Biosynthetic pathways of metabolism are usually regulated at least in part by end-product inhibition. Except for the pathways of nucleotide biosynthesis, nucleotides have no special role in the regulation of such sequences.

Strictly catabolic pathways, in contrast, do not appear to be regulated by end-product inhibition. At least in microorganisms these pathways generate substrates of energy metabolism and the activity of early enzymes of such degradative pathways is often controlled by compounds which reflect the energy state of a cell. These include nucleotides of adenine or other

purines and pyrimidines, as well as phosphate and pyrophosphate. Sanwal
(32) gives as examples, the activation by adenylate of the threonine deaminase of *E. coli* and the aspartases of *Enterobacter aerogenes* and *Bacterium
cadaveris*. Histidase in *Pseudomonas aeruginosa* is inhibited by pyrophosphate, but this is relieved by adenylate and GDP.

Amphibolic pathways, such as glycolysis, the pentose phosphate cycle,
gluconeogenesis, and the tricarboxylic acid cycle, which perform both
catabolic and anabolic roles, may be controlled both by feedback inhibition
and by indicators of energy metabolism as well as by other mechanisms.
The energy indicators of greatest interest in this regard are phosphate,
pyrophosphate, AMP, ADP, and ATP. For example, glycogen phosphorylase, phosphofructokinase, and isocitrate dehydrogenase are stimulated by
adenylate, and glycerol kinase, fructose 1,6-diphosphatase, UDP-glucose
synthetase, and β-hydroxymethylglutaryl synthetase are inhibited by
adenylate. Because of their close relationship and dependence on the
adenine nucleotides, however, other nucleotides may also serve as indicators of the energy state of the cell.

In the cases described so far, control by adenine nucleotides and other
indicators of energy status is believed to be mediated through their binding
to distinct and relatively specific sites on the enzyme in question. Such
binding results in changes in kinetic parameters, and in some cases, in the
quaternary structure of the enzyme molecule.

B. ENERGY CHARGE

Adenine and other nucleotides may also act in a regulatory fashion without having to bind to special sites on the enzymes concerned. Atkinson (33)
has proposed that many pathways of intermediary metabolism may to a
significant degree be regulated by the so-called "energy charge," which is
defined as $(ATP + 0.5\ ADP)/(ATP + ADP + AMP)$. This value varies
between 0 and 1, and is a measure of the relative concentrations of substrates and products of ATP-generating and ATP-utilizing reactions.
Feedback inhibition by end products may in some cases be superimposed
on regulation by energy charge (34). Although this hypothesis appears to
account for the regulatory behavior of adenine nucleotides in a number of
isolated enzymes, it is difficult to apply to intact cells.

References

1. Henderson, J. F., and LePage, G. A., *Chem. Rev.* **58,** 645 (1958).
2. Strominger, J. L., *Physiol. Rev.* **40,** 55 (1960).
3. Potter, V. R., "Nucleic Acid Outlines." Burgess, Minneapolis, Minnesota, 1960.

4. Mandel, P., *Progr. Nucl. Acid Res. Mol. Biol.* **3,** 299 (1964).
5. Watson, J. D., "Molecular Biology of the Gene." Benjamin, New York, 1965.
6. Lehninger, A. L., "Bioenergetics." Benjamin, New York, 1965.
7. Mahler, H. R., and Cordes, E. H., "Biological Chemistry." Harper, New York, 1966.
8. Racker, E., "Mechanisms in Bioenergetics." Academic Press, New York, 1965.
9. Torii, K., and Bandurski, R. S., *Biochim. Biophys. Acta* **136,** 286 (1967).
10. Hassid, W. Z., *Science* **165,** 137 (1969).
11. Hassid, W. Z., *Metab. Pathways* **1,** 307 (1969).
12. Dutton, G. H., ed., "Glucuronic Acid: Free and Combined, Chemistry, Biochemistry, Pharmacology, and Medicine." Academic Press, New York, 1966.
13. Straub, A. M., and Westphal, O., *Bull. Soc. Chim. Biol.* **46,** 1647 (1964).
14. Shapiro, S. K., and Schlenk, F., eds., "Transmethylation and Methionine Biosynthesis." Univ. of Chicago Press, Chicago, Illinois, 1965.
15. Goldman, P., and Vegelos, P. R., *Compr. Biochem.* **15,** 71 (1964).
16. Kornberg, A., *Science* **163,** 1410 (1969).
17. Richardson, J. P., *Progr. Nucl. Acid Res. Mol. Biol.* **9,** 75 (1969).
18. Wu, C. W., and Goldthwait, D. A., *Biochemistry* **8,** 4450 (1969).
19. Robison, G. A., Butcher, R. W., and Sutherland, E. W., *Annu. Rev. Biochem.* **27,** 149 (1968).
20. Robison, G. A., Sutherland, E. W., and Butcher, E. W., "Cyclic AMP." Academic Press, New York, 1971.
21. Kuo, J. F., and Greengard, P., *J. Biol. Chem.* **244,** 3417 (1969).
22. Skoog, F., and Armstrong, D. J., *Annu. Rev. Plant Physiol.* **21,** 359 (1970).
23. Hall, R. H., *Progr. Nucl. Acid Res. Mol. Biol.* **10,** 57 (1970).
24. Berne, R. M., *Amer. J. Physiol.* **204,** 317 (1963).
25. Berne, R. M., *in* "The Myocardial Cell" (S. A. Briller and H. L. Conn, Jr., eds.), p. 355. Univ. of Pennsylvania Press, Philadelphia, 1966.
26. Baer, H. P., and Drummond, G. I., *Proc. Soc. Exp. Biol. Med.* **127,** 33 (1968).
27. Daniel, E. E., and Irwin, J., *Can. J. Physiol. Pharmacol.* **43,** 89 (1965).
28. Michal, F., and Firkin, B. G., *Annu. Rev. Pharmacol.* **9,** 95 (1969).
29. Holmsen, H., Day, H. J., and Storm, E., *Biochim. Biophys. Acta* **186,** 254 (1969).
30. Stadtman, E. R., *Advan. Enzymol.* **28,** 41 (1966).
31. Blakeley, R. L., and Vitols, E., *Annu. Rev. Biochem.* **37,** 201 (1968).
32. Sanwal, B. D., *Bacteriol. Rev.* **34,** 20 (1970).
33. Atkinson, D. E., *Annu. Rev. Biochem.* **35,** 85 (1966).
34. Atkinson, D. E., *Biochemistry* **7,** 4030 (1968).

CHAPTER 4

THE THREE LEVELS OF PHOSPHORYLATION

I. Transphosphorylation Mechanisms

The three levels of phosphorylation in the free nucleotide pool of tissues (i.e., the nucleoside mono-, di-, and triphosphates) are interconnected by readily reversible transphosphorylation reactions:

$$\text{Nucleoside monophosphate (NMP)} \underset{(1)}{\longleftrightarrow} \text{Nucleoside diphosphate (NDP)} \underset{(2)}{\longleftrightarrow} \text{Nucleoside triphosphate (NTP)}$$

These reactions, nucleoside monophosphate kinase (reaction 1) and nucleoside diphosphate kinase (reaction 2), involve the formation and cleavage of pyrophosphoryl bonds:

$$B_1RP + B_2RP\text{-}P\text{-}P \leftrightarrow B_1RP\text{-}P + B_2RP\text{-}P \tag{1}$$

$$B_1RP\text{-}P + B_2RP\text{-}P\text{-}P \leftrightarrow B_1RP\text{-}P\text{-}P + B_2RP\text{-}P \tag{2}$$

These transfers involve only the *exchange* of high-energy phosphoryl groups: a phosphoryl group is transferred from a pyrophosphoryl linkage in a donor compound to form a new pyrophosphoryl linkage in a recipient compound. Because the breakage of each high-energy bond in the substrate results in the formation of another such bond in the product, the energy potential of the phosphoanhydride bond is retained and the reactions are freely reversible.

The facile reversibility of these transphosphorylations, and the presence in cells of a multiplicity of kinases, are apparent in an experiment by Zahn *et al.* (*1*). Inorganic phosphate labeled with ^{32}P was injected intravenously into mice and after short time intervals liver samples were taken with clamps which had been chilled in liquid nitrogen; virtually instantaneous freezing is achieved with this technique. After isolation of the indi-

TABLE 4–I

INCORPORATION OF ^{32}P-ORTHOPHOSPHATE INTO THE
ACID-SOLUBLE NUCLEOTIDES OF MOUSE LIVER
MEASURED AT SHORT TIME INTERVALS AFTER
INTRAVENOUS INJECTION[a]

Nucleotide	Post-injection interval[b]			
	10 sec	20 sec	30 sec	60 sec
ARPP*	16	41	55	71
ARPP*P	2	31	40	56
URPPP*	44	73	81	84
URPP*	10	20	—	—
URPP*P	1	4	23	38
GRPPP*	48	79	94	—
GRPP*P	1	5	14	—
ARPPP*	78	169	238	440

[a] From (*1*). Reproduced with permission.

[b] Specific activity of the indicated (*) phosphate expressed as a percentage of the specific activity of the γ-phosphate of ATP (relative to plasma P_i at 20 sec).

vidual nucleotides by ion-exchange chromatography, the specific activities of the β- and γ-phosphates of the individual nucleotides were determined. With the very short sampling times possible with this technique, progressive incorporation of isotope into the labile phosphates of the liver nucleotides was observed, as may be seen in Table 4–I; to simplify the interpretation of these results, it may be reasonably assumed that the γ-P of ATP was the principal isotopic donor for the observed ^{32}P-incorporations.

Similar results had been obtained earlier by Brumm *et al.* (*2*) using rat liver; they demonstrated that the specific activity of the α-phosphates of the liver nucleotides had not reached that of inorganic phosphate by 60 minutes after isotope administration.

This facile distribution of phosphate among the high-energy phosphates of the four ribonucleoside phosphate families may be explained by these processes:

(a) Conversion of ADP to ATP by mitochondrial oxidative phosphorylation;

(b) Enzymatic transphosphorylations among the adenosine phosphates;

(c) Transphosphorylations between the adenosine phosphates and other nucleoside phosphates.

The ready reversibility of the kinases and their abundance in cells has the result that all of the free nucleoside phosphates in a cell reflect the relative amounts of the adenosine phosphates in each of the three levels of phosphorylation. Because the adenosine phosphates are primarily involved in energy metabolism, the energy status of the cell is reflected in the relative amounts of ATP, ADP, and adenylate. Through the very facile transphosphorylation reactions, other nucleoside phosphates become similarly distributed among the three levels of phosphorylation; perturbations in energy metabolism, such as those caused by substrate deprivation or by anoxia, cause parallel shifts in the relative amounts of the mono-, di-, and triphosphates in all of the nucleotide families.

This discussion attempts to establish the means by which the three levels of phosphorylation in the nucleotide pool are interconnected; the input of high-energy phosphate bonds into this pool by substrate-level phosphorylations and by oxidative phosphorylation is mentioned in Chapter 3. Certain kinases are part of the chemical machinery of the mitochondrion, but these will be considered only to the extent that they serve as connections between the mitochondrial generation of phosphoanhydride groups and the nucleotide pool. Utilization of phosphoanhydride bond energy in synthetic reactions, and the consumption of the nucleotides themselves for synthesis of coenzymes and polynucleotides, will, of course, influence the distribution of the nucleotide pool constituents among the three levels of phosphorylation.

II. Nucleoside Monophosphate Kinases

These enzymes catalyze the reversible transfer of phosphoryl groups between nucleoside triphosphates and nucleoside monophosphates:

$$N_1TP + N_2MP \rightleftharpoons N_1DP + N_2DP$$

The first such enzymatic transphosphorylation to be studied was the formation of ADP from adenylate and ATP, catalyzed by an enzyme in rabbit muscle (3); this enzyme, "myokinase," is one of a larger group of adenylate kinases. It is apparent today that the nucleoside monophosphate kinases constitute a large family of enzymes with different specificities for nucleotide substrates; certain of these kinases have particular intracellular sites (see reference 4).

Because the majority of the known nucleoside monophosphate kinase reactions require an adenosine phosphate as one of the substrates, they may be classified as those which (1) are specific for adenylate as a phosphoryl acceptor, and (2) as those which require ATP as a phosphoryl donor. (The adenylate kinase reaction obviously may be placed in either category.) There may exist an additional class of nucleoside monophosphate kinase reactions in which adenosine phosphates do not participate; however, such enzyme activity has not been unequivocally demonstrated.

A. ATP:NUCLEOSIDE MONOPHOSPHATE KINASES

1. *Adenylate Kinase*

Table 4–II shows the adenylate kinase activities of unfractionated soluble extracts of several rat tissues (5). The assay used was ADP synthesis from ATP and adenylate; these data indicate that rat tissues differ widely in their adenylate kinase activities. The adenylate kinase activity of liver cells is also present in mitochondria as well as in the soluble phase but is absent from nuclei and microsomes; about 90% of the total activity in rat liver is found in the cytosol. The mitochondrial enzyme appears not to be a contaminant from the cytosol.

Adelman et al. (5) have shown that the soluble adenylate kinase activity of liver is strikingly responsive to dietary alteration. Fasting for 48 hours resulted in a three-fold increase in the concentration of the soluble adenylate kinase activity in rat liver, with no attendant changes in the ac-

TABLE 4–II

OCCURRENCE OF SOLUBLE "ADENYLATE KINASE"
ACTIVITY IN TISSUES OF THE RAT[a]

Tissue	Activity	
	Units/g tissue	Units/mg protein
Liver	135	2.4
Kidney	80	1.2
Intestinal mucosa	14	0.005
Adipose tissue	3	0.66
Brain	13	0.38
Heart muscle	<0.1	—
Skeletal muscle	<0.1	—

[a] From (5). Reprinted with permission from R. C. Adelman, C. H. Lo, and S. Weinhouse, *Advan. Enzyme Regul.* **6**, 425 (1968), Pergamon Press Ltd.

tivity of the mitochondrial adenylate kinase. The elevated levels of the soluble enzyme in fasted animals were restored dramatically to prefasting levels by glucose feeding; the enzyme activity is not repressed by glucose per se, but insulin may be involved, because insulin administration produces a profound fall in the level of the adenylate kinase activity in the livers of fed, diabetic rats.

The physiological significance of the responsiveness of this enzyme to glucose and insulin is not yet established; however, it may be related to gluconeogenesis, because several other key enzymes in this process show similar responses.

Two preparations of adenylate kinase have been highly purified (see reference 4):

a. *Muscle Adenylate Kinase* ("*Myokinase*"). This enzyme has unusual stability to heat and acid. The homogenous enzyme has a molecular weight of 21,000 and has two sulfhydryl groups that are essential to activity. The substrate specificity is high, with the participation of CDP in the reverse reaction being virtually the only known departure from the requirement for adenosine phosphates. The enzyme requires Mg^{2+} and it appears that the substrates are AMP^{2-} and $MgATP^{2-}$.

Recently, Rhoads and Lowenstein (6) have utilized initial velocity and equilibrium kinetics to study the kinetic mechanism of muscle adenylate kinase from rabbit muscle and have concluded that the reaction is of the random bi-bi type. Isotope exchange between adenylate and ADP is faster than that between ATP and ADP. Apparently, the rate-limiting steps in the reaction include dissociation of the nucleotides from the enzyme.

The two sulfhydryl groups of muscle adenylate kinase are not involved in substrate binding, although the enzyme may be reversibly inactivated with sulfhydryl reagents (7, 8).

b. Mitochondrial Adenylate Kinase. This enzyme activity has been isolated from bovine liver mitochondria (9). This enzyme would appear to function in the conversion of extramitochondrial ADP to ATP and probably does not have an obligatory role in oxidative phosphorylation. The mitochondrial adenylate kinase differs from muscle adenylate kinases in having no free sulfhydryl groups, and it is markedly less stable to heat and acid. This enzyme has a high specificity, showing no reaction with mono- and diphosphates of inosine, uridine, and cytidine; however, the adenine deoxyribonucleotides are substrates for the mitochondrial enzyme. The divalent metal ion requirement and molecular weights are similar to those for muscle adenylate kinase.

2. *Guanylate Kinase*

Guanylate kinase has been partly purified from hog brain (10), from *Escherichia coli* (11), and from a transplantable mouse tumor, Sarcoma 180 (12). These preparations show similarities with respect to substrate specificity and cation requirements, but differ with respect to pH optima and molecular weight. All three preparations are highly specific in that guanylate and deoxyguanylate are phosphorylated, but adenylate, inosinate, xanthylate, and cytidylate are not substrates. The brain and tumor enzymes also phosphorylate 8-azaguanylate. ATP or dATP are the specific phosphate donors for the three enzymes.

Meich and Parks (10) have shown that in the transphosphorylation reaction with the brain enzyme, both substrates are bound before the product is released and a phosphorylated enzyme intermediate is probably not involved in the reaction.

The acceptability of deoxyguanylate and dATP as substrates in the guanylate kinase reaction is noteworthy; there does not appear to be a particular group of nucleoside monophosphate kinases specific for deoxyribonucleotides (see Chapter 14), apart from that for thymidylate and deoxyuridylate.

3. *Uridylate-Cytidylate Kinase*

In a preparation from calf liver, Strominger *et al.* (13) achieved a partial separation of uridylate-cytidylate kinase activity from those for guanylate and adenylate. Maley and Ochoa (14) partly purified a kinase from *Azoto-bacter vinelandii* that catalyzed phosphoryl transfer from ATP to deoxy-cytidylate and cytidylate, but not to adenylate, inosinate, guanylate, or deoxyguanylate. More recently, Ruffner and Anderson (15) have purified 300-fold an enzyme from *Tetrahymena pyriformis* that catalyzes the phosphorylation of uridylate, cytidylate, and deoxycytidylate, but not that of adenylate, guanylate, deoxyguanylate, or thymidylate; of various phosphate donors tested, only ATP and dATP were substrates. Parallel behavior of the uridylate and cytidylate kinase activities during the isolation procedure indicated that both activities are catalyzed by the same enzyme; however, Hiraga and Sugino (16) appear to have separated the uridylate and cytidylate-deoxycytidylate kinase activities of *E. coli*. The *T. pyriformis* enzyme is inhibited by CDP, UDP, and ADP, findings which Ruffner and Anderson interpret in terms of the regulation of pyrimidine nucleotide biosynthesis in this pyrimidine-requiring organism.

Sugino *et al.* (17) have partly purified a deoxycytidylate kinase from calf thymus; this preparation also phosphorylates cytidylate and uridylate, but not deoxyuridylate. The deoxyuridylate kinase activity was recovered in another protein fraction and so represents a distinct entity. As phosphate donors, only ATP and dATP were active in the cytidylate reaction.

Thus, while existence of a cytidylate-deoxycytidylate kinase is established, there remains some uncertainty about the association of uridylate kinase activity with this enzyme. Phosphorylation of deoxyuridylate appears to be catalyzed by a separate enzyme activity, possibly by thymidylate kinase.

4. Thymidylate Kinase

Nelson and Carter (18) have purified thymidylate kinase about 5,000-fold from *E. coli* B and have shown that thymidylate was phosphorylated about seven times faster than deoxyuridylate. Uridylate was not a substrate, nor were adenylate, guanylate, cytidylate, or their deoxyribosyl counterparts. This enzyme was less specific for the triphosphate phosphoryl donor than the kinases previously discussed, because CTP, GTP, and the corresponding deoxyribonucleotides also served as phosphate donors, although less well than ATP.

5. Deoxynucleoside Monophosphate Kinases

It appears that separate phosphokinases do not exist for the deoxy-ribonucleoside monophosphates of adenine, guanine, and cytosine, and that phosphorylation of these compounds is accomplished by the enzyme specific for the corresponding ribonucleotide (see Chapter 14). However, the identity of the enzymes responsible for phosphorylation of uridylate and deoxyuridylate is uncertain. Thymidylate phosphorylation, in *E. coli* at least, is catalyzed by a particular kinase with a specificity somewhat broader than that of the other nucleoside monophosphate kinases that have been purified; this enzyme can accept deoxyuridylate as a substrate.

6. *Phosphorylation of Inosinate*

The nucleoside monophosphate kinases do not appear to catalyze the phosphorylation of inosinate. Hershko *et al.* (*19*) did not detect inosinate phosphorylation in stroma-free hemolysates of human erythrocytes.

B. NUCLEOSIDE TRIPHOSPHATE: ADENYLATE KINASES

Nucleoside monophosphate kinases have been isolated which are specific for adenylate as the phosphoryl acceptor, but will accept various nucleoside triphosphates as phosphoryl donors. Such an enzyme activity was partly purified from hog kidney by Gibson *et al.* (*20*). Heppel *et al.* (*21*) have separated a phosphokinase activity from calf liver which accepts only adenylate as the nucleoside monophosphate substrate, but uses ATP, GTP, CTP, UTP, or ITP as the phosphate donor.

The GTP-adenylate kinase isolated from swine liver by Chiga *et al.* (*22*) has a substrate specificity narrower than the above activities (which may possibly represent more than one enzyme). The GTP:adenylate kinase is specific for adenylate (deoxyadenylate), but will accept ITP as a phosphoryl donor in place of GTP. This enzyme is localized in the mitochondrion and is thought to link the adenosine phosphates with substrate level phosphorylation in the oxidation of α-ketoglutarate (*23, 24*). The α-ketoglutarate dehydrogenase and succinyl coenzyme A synthetase reactions which, in concert, phosphorylate GDP, are located in the matrix and inner membrane of the mitochondrion; GTP: adenylate kinase is also located at this site (*24*) and could serve to transfer phosphoryl groups from the GTP so generated to ADP and thus to regenerate GDP. Part of the mitochondrial nucleoside diphosphate kinase also appears to be located in the inner membrane, and this enzyme may also participate in the regeneration of GDP in this mitochondrial compartment.

III. Nucleoside Diphosphate Kinases

The very active nucleoside diphosphate kinase activity of cells has been demonstrated many times (see reference *25*). The general reaction catalyzed is

$$N_1TP + N_2DP \rightleftharpoons N_1DP + N_2TP$$

This enzyme activity is widely distributed in nature, and tissues generally show a higher concentration of this activity than of nucleoside monophosphate kinases. The enzyme has broad specificity; purified preparations appear to have only the requirement that the substrates be a nucleoside triphosphate and a nucleoside diphosphate.

Nucleoside diphosphate kinase has been prepared in highly purified form from various sources; for example, from human erythrocytes by Mourad and Parks (26) and from beef heart mitochondria by Colomb et al. (27). The molecular weights of the nucleoside diphosphate kinases from human erythrocytes, yeast, bovine liver, and heart mitochondria are all in the range 100,000 to 110,000. These enzymes possess essential thiols.

The nucleoside diphosphate kinase from human erythrocytes was of low specificity and would react with nucleoside triphosphates and nucleoside diphosphates which contained either ribose or deoxyribose and any of the natural purine or pyrimidine bases. The finding that alternative substrates, such as GTP and ATP, were competitive inhibitors, is consistent with the low substrate specificity (26).

Nucleoside diphosphate kinases exhibit activity over a wide pH range, but usually have optima at or near pH 7. They require the presence of divalent metals for activity, but have a low specificity in this respect, since Mg^{2+}, Mn^{2+}, Ca^{2+}, Co^{2+}, and to a lesser extent, Ni^{2+} and Zn^{2+}, may satisfy this requirement. Magnesium is evidently the physiological ion serving this enzyme's requirement for a divalent ion. The function of magnesium ion in the nucleoside diphosphate kinase reaction is apparent in the work of Colomb et al. (27), who have shown that for the enzyme from beef heart mitochondria, MgATP, but not free ATP, serves as the phosphate donor. Further, free ADP was shown to be preferred over MgADP as the phosphate acceptor.

Mourad and Parks (26) noted, as had others, that the nucleoside diphosphate kinase activity of tissues is much higher than that of some other enzymes of nucleotide metabolism; for example, the activity of this enzyme in various hog tissues was found to be 10–30 times higher than that of guanylate kinase. The nucleoside diphosphate kinase activity of erythrocytes was also found to be high, being sufficient to phosphorylate 30 μmoles of ADP per hour per milliliter packed cells (the total nucleotide pool of erythrocytes is about 1.2 μmoles per milliliter cells).

The reaction mechanism of the erythrocyte nucleoside diphosphate kinase, on the basis of isotope exchange data, was shown to be of the "ping-pong" type. A mechanism of this nature requires the formation of an enzyme-phosphate intermediate; evidence for such an intermediate was obtained by Mourad and Parks (28). In their experiments, the purified enzyme was incubated with γ-^{32}P-ATP in the absence of nucleoside diphosphate substrate; the enzyme was isolated by chromatography on Sephadex G-200 columns and it was found to have been labeled with ^{32}P. When a nucleoside diphosphate and the isolated, labeled enzyme were incubated together, the label disappeared from the enzyme and appeared in a nucleoside triphosphate product. Other workers have also provided evidence for the formation of a phosphorylated enzyme intermediate in the reaction of nucleoside diphosphate kinases from other various sources [for example, see Pedersen (29)].

Norman et al. (30) showed that purified nucleoside diphosphate kinase from the Jerusalem artichoke was phosphorylated during incubation with ^{32}P-ATP, and from an alkaline hydrolysate of the labeled enzyme 3-phosphohistidine was isolated. In similar experiments, nucleoside diphosphate kinases from human erythrocytes and bovine liver were phosphorylated with ^{32}P-ATP and upon hydrolysis, the radioactive degradation products from the two enzymes were found to be the same and to include two phosphorylated derivatives of histidine and one of lysine, indicating that the nucleoside diphosphate kinases have similar structures at their active sites.

In summary, the nucleoside diphosphate kinase reaction is catalyzed by enzymes with broad substrate specificity; many tissues show high activities of this enzyme, which is widely distributed in nature.

IV. Kinases and the Generation of High-Energy Phosphoryl Groups

The nucleotide kinases apparently serve to distribute throughout the free nucleotide pool the high-energy phosphoryl groups generated by mitochondrial oxidative phosphorylation and substrate level phosphorylations.

The enzymatic machinery of the tricarboxylic acid cycle is located in a mitochondrial membrane system; as mentioned above, the presence of nucleoside monophosphate and diphosphate kinases in this cell compartment enables the substrate-level phosphorylation associated with α-ketoglutarate oxidation to be coupled with the adenosine phosphates.

Extramitochondrial adenosine di- and triphosphates "communicate" in a rapid and extensive two-way flux with the intramitochondrial aden-

TABLE 4–III

ENZYMES OF NUCLEOTIDE TRANSPHOSPHORYLATION

Enzyme Commission number	Systematic name	Trivial name
2.7.4.3	ATP:AMP phosphotransferase	Adenylate kinase
2.7.4.6	ATP:nucleosidediphosphate phosphotransferase	Nucleosidediphosphate kinase
2.7.4.8	ATP:GMP phosphotransferase	Guanylate kinase
2.7.4.9	ATP:thymidinemonophosphate phosphotransferase	Thymidylate kinase
2.7.4.10	GTP:AMP phosphotransferase	GTP-adenylate kinase Uridylate-cytidylate kinase Uridylate kinase Cytidylate-deoxy-cytidylate kinase

ine nucleotides, by means of a specific "carrier" mechanism located in the outer mitochondrial membrane. Other extramitochondrial nucleotides are not accepted by the carrier and, therefore, do not have access to the interior of the mitochondrion. Adenine nucleotides, charged with high-energy phosphoryl groups generated by oxidative phosphorylation, pass out of the mitochondrion by the carrier ("translocase"). These phosphoryl groups are transferred to other members of the nucleotide pool by the kinases considered above. The translocation of adenosine phosphates across the mitochondrial membrane is inhibited by atractyloside, a perhydrophenthrenic glycoside from the thistle, *Atractylis gummifera*.

The enzymes of nucleotide transphosphorylation are listed in Table 4–III.

References

1. Zahn, D., Klinger, R., and Frunder, H., *Eur. J. Biochem.* 11, 549 (1969).
2. Brumm, A. F., Potter, V. R., and Siekevitz, P., *J. Biol. Chem.* 220, 713 (1956).
3. Colowick, S. P., and Kalckar, H. M., *J. Biol. Chem.* 148, 117 (1943).
4. Noda, L., *in* "The Enzymes" (P. D. Boyer, H. Lardy, and K. Myrbäck, eds.), 2nd rev. ed., Vol. 6, p. 139. Academic Press, New York, 1962.
5. Adelman, R. C., Lo, C. H., and Weinhouse, S., *Advan. Enzyme Regul.* 6, 425 (1968).
6. Rhoads, D. G., and Lowenstein, J. M., *J. Biol. Chem.* 243, 3963 (1968).
7. Kress, L. F., Bono, V. H., Jr., and Noda, L., *J. Biol. Chem.* 241, 2293 (1966).
8. Kress, L. F., and Noda, L., *J. Biol. Chem.* 242, 558 (1967).

9. Markland, F. S., and Wadkins, C. L., *J. Biol. Chem.* **241**, 4136 (1966).
10. Miech, R. P., and Parks, R. E., Jr., *J. Biol. Chem.* **240**, 351 (1965).
11. Oeschger, M. P., and Bessman, M. J., *J. Biol. Chem.* **241**, 5452 (1966).
12. Miech, R. P., York, R., and Parks, R. E., Jr., *Mol. Pharmacol.* **5**, 30 (1969).
13. Strominger, J. L., Heppel, L. A., and Maxwell, E. S., *Biochim. Biophys. Acta* **32**, 412 (1959).
14. Maley, F., and Ochoa, S., *J. Biol. Chem.* **233**, 1538 (1958).
15. Ruffner, B. W., Jr., and Anderson, E. P., *J. Biol. Chem.* **244**, 5994 (1969).
16. Hiraga, S., and Sugino, Y., *Biochim. Biophys. Acta* **114**, 416 (1966).
17. Sugino, Y., Teraoka, H., and Shimono, H., *J. Biol. Chem.* **241**, 961 (1966).
18. Nelson, D. J., and Carter, C. E., *J. Biol. Chem.* **244**, 5254 (1969).
19. Hershko, A., Razin, A., Shoshani, T., and Mager, J., *Biochim. Biophys. Acta* **149**, 59 (1967).
20. Gibson, D. M., Ayengar, P., and Sanadi, D. R., *Biochim. Biophys. Acta* **21**, 86 (1956).
21. Heppel, L. A., Strominger, J. L., and Maxwell, E. S., *Biochim. Biophys. Acta* **32**, 422 (1959).
22. Chiga, M., Rogers, A. E., and Plaut, G. W. E., *J. Biol. Chem.* **236**, 1800 (1961).
23. Heldt, H. W., and Schwalbach, K., *Eur. J. Biochem.* **1**, 199 (1967).
24. Lima, M. S., and Vignais, P. V., *Bull. Soc. Chim. Biol.* **50**, 1833 (1968).
25. Weaver, R. H., *in* "The Enzymes" (P. D. Boyer, H. Lardy, and K. Myrbäck, eds.), 2nd rev. ed., Vol. 6, p. 151. Academic Press, New York, 1962.
26. Mourad, N., and Parks, R. E., Jr., *J. Biol. Chem.* **241**, 271 (1966).
27. Colomb, M. G., Cheruy, A., and Vignais, P. V., *Biochemistry* **8**, 1926 (1969).
28. Mourad, N., and Parks, R. E., Jr., *Biochem. Biophys. Res. Commun.* **19**, 312 (1965).
29. Pedersen, P. L., *J. Biol. Chem.* **243**, 4305 (1968).
30. Norman, A. W., Wedding, R. T., and Black, M. K., *Biochem. Biophys. Res. Commun.* **20**, 703 (1965).

CHAPTER 5

CARBON AND NITROGEN TRANSFER REACTIONS IN NUCLEOTIDE METABOLISM

In addition to the transphosphorylation reactions discussed in Chapter 4, there are several general types of carbon and nitrogen transfer reactions which also occur in purine and pyrimidine nucleotide biosynthesis and interconversion. Among these are one-carbon and phosphoribosyl transfer reactions, amino group transfer from glutamine and aspartate, and amide syntheses. In most of these processes carbon–nitrogen bonds are formed, although some of these processes involve the synthesis of carbon–carbon bonds. In this chapter these reaction types will be discussed in a general fashion to provide background for the detailed treatment of individual examples in later chapters.

The carbon and nitrogen transfer reactions discussed here in the context of nucleotide metabolism are also common in amino acid metabolism and illustrate well the close relationship of these two areas. They are all discussed in Meister's study of amino acids (1).

I. Folic Acid Coenzyme Reactions

The enzymatic roles of the folate coenzymes were first studied in connection with purine nucleotide biosynthesis, although some of the major manifestations of folate deficiency are due to impairment of thymidylate and DNA synthesis. Figure 5-1 shows the structure of folic acid and identifies its component parts. Nomenclature in this area has changed frequently and is sometimes confusing and inconsistent; Table 5–I gives the latest approved nomenclature, together with alternatives also found in the literature. The field of folic acid biochemistry is discussed in depth by Blakley (2).

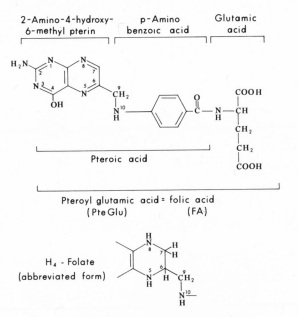

FIG. 5-1. Folic acid structure and nomenclature.

TABLE 5–I

NOMENCLATURE OF FOLATE COENZYMES

A. One-carbon groups
   –CH_3 Methyl
   –CH_2–   Methylene (formerly "hydroxymethyl")
   –CHO   Formyl
   –CHNH   Formimino, formimidoyl
   –CH =   Methenyl, methylidyne
B. Synonyms for H_4-folate coenzymes
         5-Formyl H_4-folate (folinic acid, leucovorin,
                                         citrovorum factor)
         5,10-Methenyl H_4-folate (anhydroleucovorin)
         5-Methyl H_4-folate (prefolic A)

A. INTERCONVERSION OF FOLATE COENZYMES

Reduced folate coenzymes transfer one-carbon moieties at several oxidation levels and in several different ways. These coenzyme forms are, for the most past, interconvertible, as indicated in Fig. 5-2.

B. REACTIONS

Table 5–II lists the major reactions in which folate coenzymes are known to participate, presented according to the oxidation level of the one-carbon fragment transferred. In addition, folate coenzymes are known or suspected to participate in several other processes, but their details are still unclear (2).

C. ANTIMETABOLITES

The folic acid analogues, methotrexate (amethopterin) and aminopterin, prevent the utilization of folic acid and of one-carbon units, and produce in animals symptoms of deficiency of this vitamin.

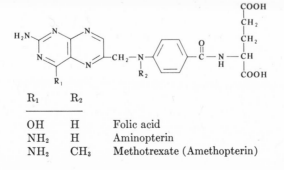

R_1	R_2	
OH	H	Folic acid
NH_2	H	Aminopterin
NH_2	CH_3	Methotrexate (Amethopterin)

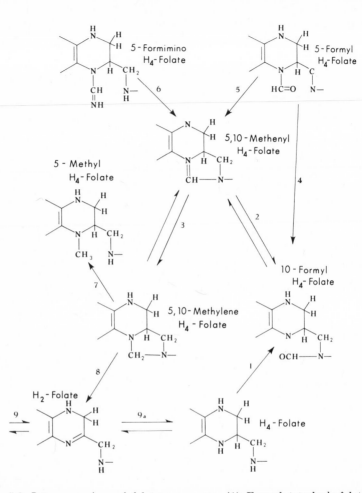

Fig 5-2. Interconversions of folate coenzymes. (1) Formyl tetrahydrofolate synthetase, 6.3.4.3; (2) methenyl tetrahydrofolate cyclohydrolase, 3.5.4.9; (3) methylene tetrahydrofolate dehydrogenase, 1.5.1.5; (4) 5-formyl tetrahydrofolate isomerase; (5) 5-formyl tetrahydrofolate isomerase (cyclodehydrase); (6) formimino tetrahydrofolate cyclodeaminase, 4.3.1.4; (7) 5,10-methylene tetrahydrofolate reductase, 1.1.1.68; (8) thymidylate synthetase; (9) dihydrofolate dehydrogenase, 1.5.1.4; (9a) tetrahydrofolate dehydrogenase, 1.5.1.3.

The folate analogues are powerful inhibitors of tetrahydrofolate dehydrogenase (or folate reductase), the enzyme which converts folate to the active tetrahydro derivative. Reduced forms of aminopterin also inhibit tetrahydrofolate dehydrogenase, dihydroaminopterin being about as

TABLE 5–II

REACTIONS INVOLVING FOLATE COENZYMES

Oxidation level	Reaction	Enzyme Commission number
Formate	Phosphoribosyl glycineamide → phosphoribosyl formylglycineamide + H_4-folate + 5,10-methenyl H_4-folate	2.1.2.2
	Phosphoribosyl aminoimidazole carboxamide → phosphoribosyl formimidoimidazole + 10-formyl H_4-folate carboxamide + H_4-folate	2.1.2.3
	Formiminoglycine + H_4-folate → glycine + 5-formimino H_4-folate	2.1.2.4
	Formiminoglutamate + H_4-folate → glutamate + 5-formimino H_4-folate	2.1.2.5
	Formylglutamate + H_4-folate → glutamate + 5-formyl H_4-folate	2.1.2.6
Formaldehyde	Deoxyuridylate + 5,10-methylene H_4-folate → thymidylate + H_2-folate	—
	Deoxycytidylate + 5,10-methylene H_4-folate → 5-hydroxymethyl-deoxycytidylate + H_4-folate	—
Methyl	Serine + H_4-folate → glycine + 5,10-methylene H_4-folate	2.1.2.1
	5-Methyl H_4-folate + homocysteine → methionine + H_4-folate	—

effective as aminopterin, but considerably more potent than tetrahydro-aminopterin (3).

The affinity of the antagonists for tetrahydrofolate dehydrogenase is quite remarkable, with reported binding constants in the order of 10^{-9} M. The binding of the inhibitor to the dehydrogenase, although exceedingly firm, is still reversible by procedures such as dialysis against a high concentration of folic acid, or by passage of the dehydrogenase–analogue complex through columns of hydroxylapatite or DEAE-cellulose (3).

Methotrexate has an important clinical use as an antitumor agent. Some patients with acute leukemia respond to methotrexate treatment with remission of their disease, but such responses are usually transitory. Spectacular responses to methotrexate treatment are achieved in the childhood neoplasm, Burkitt's lymphoma, and in choriocarcinoma; in both, long-term responses are obtained. In what is probably its most important current application, methotrexate is used in various combinations with other anticancer drugs in the treatment of human leukemia.

II. Glutamine Amide Transfer Reactions

A. REACTIONS

Table 5–III lists the glutamine amide transfer reactions of purine and pyrimidine biosynthesis *de novo*, and of purine and pyrimidine ribonucleotide interconversion, and several more that occur in other areas of metabolism. All have features in common (1, 3).

B. MECHANISMS

In most cases, NH_3 is able to replace glutamine as the nitrogen donor; the Michaelis constants for both substrates are similar. However, it is NH_3 rather than NH_4^+ which participates in these reactions, so that at pH 7.4 the total ($NH_3 + NH_4^+$) concentration must be about 100 times that of glutamine to give the same reaction rate. It would appear, therefore, that in these reactions glutamine brings to the enzyme an uncharged nitrogen atom at the oxidation level of ammonia. The physiological importance of the replacement of glutamine by NH_3 appears to vary from enzyme to enzyme, and even from organism to organism for the same enzyme. In most cases hydroxylamine can replace ammonia as substrate, the corresponding hydroxylamino derivative being a product.

In the cases studied, the ammonia or glutamine reacts with an activated, enzyme-bound acceptor. Thus Meister (4) has suggested that the glutamine-dependent carbamyl phosphate synthetase reaction may follow

TABLE 5–III

GLUTAMINE AMIDE TRANSFER REACTIONS

Amide group acceptor	Product	Enzyme Commission number
Phosphoribosyl pyrophosphate	Ribosylamine phosphate	2.4.2.14
Phosphoribosyl formyl-glycineamide	Phosphoribosyl formyl-glycineamidine	6.3.5.2
Xanthylate	Guanylate	6.3.5.3 6.3.4.1
Uridine triphosphate	Cytidine triphosphate	6.3.4.2
Deamido-nicotinamide adenine dinucleotide	Nicotinamide adenine dinucleotide	6.3.5.1
Phosphodeoxyribulosyl formimino-phosphoribosyl aminoimidazole carboxamide	Phosphodeoxyribulosyl amidino-phosphoribosyl aminoimidazole carboxamide	—
Fructose 6-phosphate	Glucosamine 6-phosphate	—
CO_2 + ATP	Carbamyl phosphate	—
Aspartate	Asparagine	—
Chorismate	o-Aminobenzoate	—
Chorismate	p-Aminobenzoate	—

this sequence:

$$Enzyme + ATP + HCO_3^- \rightarrow enzyme\text{-}CO\text{-}P + ADP$$

$$Enzyme\text{-}CO\text{-}P + Gln \rightarrow enzyme\text{-}CONH_2 + Glu + P_i$$

$$Enzyme\text{-}CONH_2 + ATP \rightarrow H_2NCOP + ADP + enzyme$$

and asparagine synthesis, may follow this route:

$$Enzyme + Asp + ATP \rightarrow enzyme\text{-}Asp\text{-}AMP + PP_i$$

$$Enzyme\text{-}Asp\text{-}AMP + Gln \rightarrow enzyme + Aspn + AMP + Glu$$

As will be discussed in Chapter 8, the following sequence is followed by guanylate synthetase:

$$Enzyme + XMP + ATP \rightarrow enzyme\text{-}XMP\text{-}AMP + PP_i$$

$$Enzyme\text{-}XMP\text{-}AMP + NH_3 \rightarrow GMP + AMP + enzyme$$

The two amidotransferases in the purine biosynthetic pathway differ in that one requires ATP while the other does not. It would appear that the pyrophosphate–glycoside bond of PP-ribose-P is of the high-energy type, and that PP-ribose-P serves both as substrate and energy source in the first reaction in this pathway. In the synthesis of phosphoribosyl formylglycineamidine, physical (5, 6) and kinetic (7) studies suggest the involvement of enzyme-glutamyl and enzyme-phosphate intermediates.

C. ANTIMETABOLITES

Another common feature of the amidotransferases is that they are inhibited by the antibiotics azaserine (O-diazoacetyl-L-serine) and 6-diazo-5-oxo-L-norleucine when glutamine is substrate. However, these compounds are much less inhibitory when NH_3 is the nitrogen donor. Recently two new glutamine analogues have been introduced, 2-amino-4-oxo-5-chloropentanoic acid (2) and albizziin (8). The structures of these compounds are given in Fig. 5-3.

The ratio of the Michaelis constant of glutamine to inhibitor constants for azaserine and diazo-oxo-norleucine for the two amidotransferases of purine biosynthesis in pigeon liver are given in the accompanying tabulation.

Michaelis constant/Inhibitor constant		
	Azaserine	Diazo oxo-norleucine
PP-Ribose-P + glutamine	0.25	680
Phosphoribosyl formylglycineamide + glutamine	18	730

It may be seen that although diazo-oxo-norleucine is a much more potent inhibitor than azaserine, it is less selective. In contrast, azaserine is 60 times more effective than diazo-oxo-norleucine toward phosphoribosyl formylglycineamidine synthetase. These ratios are markedly different when determined with the *Salmonella* enzyme, however (9). Azaserine, and presumably diazo-oxo-norleucine, bind covalently to a cysteine sulfhydryl of the enzyme, with the loss of the diazo group (10).

III. Aspartate Amino Transfer Reactions and Amide Syntheses

A. REACTIONS

Transfer of the amino group of aspartate is a second method of providing amino nitrogen in synthetic reactions without the involvement of

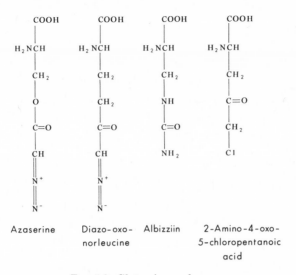

Fig. 5-3. Glutamine analogues.

free ammonia (*1*). In contrast to the glutamine amide transfer reactions, the donated amino group in this case is protonated. The products formed in these reactions are amides, amidines, or heteroaromatic amines which may be thought of as cyclic amidines.

The two reactions of this type in purine metabolism are listed in Table 5–IV, together with a similar reaction, argininosuccinate synthesis.

Another reaction of purine biosynthesis *de novo* bears a resemblance to these reactions: the synthesis of the amide, phosphoribosyl glycineamide. This is also listed in Table 5–IV, together with two closely related reactions, the synthesis of glutamine and of glutathione.

B. Mechanisms

The six reactions listed in Table 5–IV are similar in that (1) protonated ammonia or a carrier of protonated ammonia is involved; (2) all are reversible; (3) no isotope exchange or reaction occurs in either direction unless all three reactants are present; (4) all require ATP or GTP; (5) ^{18}O from phosphate is transferred to the amino acceptor. (The case of argininosuccinate synthesis is different in that pyrophosphate is formed, and ^{18}O from citrulline is found in adenylate rather than in pyrophosphate.)

In his studies of the reactions of purine biosynthesis, Buchanan suggested that reactions which involve two substrates plus ATP be called

TABLE 5–IV

TRANSFER OF THE AMINO NITROGEN OF ASPARTATE AND
AMIDE SYNTHESES

Substrates	Products	Enzyme Commission number
A. Transfer of the amino nitrogen of aspartate		
Aspartate, ATP, phosphoribosyl amino-imidazole carboxylate	ADP, P_i, phosphoribosyl amino-imidazole succino-carboxamide	6.3.2.6
Aspartate, GTP, inosinate	GDP, P_i, adenylosuccinate	6.3.4.4
Aspartate, ATP, citrulline	AMP, PP_i, argininosuccinate	6.3.4.5
B. Amide syntheses		
Glycine, ATP, ribosylamine phosphate	ADP, P_i, phosphoribosyl glycineamide	6.3.1.3
NH_3, ATP, glutamate	ADP, P_i, glutamine	6.3.1.2
Glycine, ATP, glutamylcysteine	ADP, P_i, glutathione	6.3.2.3

"kinosynthases," and proposed a concerted mechanism of this kind:

$$\text{Adenine-ribose} - \overset{\overset{O}{\|}}{\underset{\underset{O_-}{|}}{P}} - O \overset{+}{-} \overset{\overset{O^-}{|}}{\underset{\underset{O_-}{|}}{P}} - O --- \overset{\overset{O^- O^-}{\diagdown \diagup}}{\underset{\underset{O_-}{|}}{P}} --- O --- \overset{\overset{O}{\|}}{\underset{\underset{+NH_3}{\underset{|}{RCH}}}{C}} ---- \overset{\overset{H}{\vdots}}{\underset{\underset{H}{|}}{N}} - R$$

Meister has more recently studied the glutamine synthetase reaction in considerable detail, and has proposed the following mechanism:

$$\text{Enzyme} \xrightarrow{\quad ATP \quad} \text{Enzyme} \cdot \text{ATP} \dashrightarrow \text{Enzyme} \cdot \text{ATP} \cdot \text{Glu} \xrightarrow{\quad Glu \quad} \text{Enzyme} \cdot \text{ADP} \cdot \text{Glu}\,\gamma\,\text{P}$$

$$\xrightarrow{\quad NH_3 \quad} \text{Enzyme} \cdot \text{ADP} \cdot \text{Gln} \dashrightarrow \text{Enzyme} + \text{ADP} + \text{Gln}$$

The formation of enzyme-bound γ-glutamyl phosphate could explain the ^{18}O transfer data, and the mandatory addition of ATP before addition of glutamate could explain why no isotope exchange was detected with less than a complete system.

It is still not possible to decide which type of mechanism is operative for the reactions of purine metabolism. This type of reaction is now called "ligase" rather than "kinosynthase."

IV. Phosphoribosyl Transfer Reactions

A. REACTIONS

The 5-phosphoribosyl group is transferred from phosphoribosyl pyrophosphate to a wide variety of nitrogenous compounds in several areas of metabolism; these reactions are listed in Table 5–V.

TABLE 5–V

PHOSPHORIBOSYL TRANSFER REACTIONS

Phosphoribosyl acceptor	Product	Enzyme Commission number
Adenine	Adenylate	2.4.2.7
Guanine, hypoxanthine	Guanylate, inosinate	2.4.2.8
Xanthine	Xanthylate	—
Glutamine, NH_3	Ribosylamine phosphate	2.4.2.14
Orotate	Orotidylate	2.4.2.10
Uracil, 2,4-Dioxypyrimidines; 2,6-diketopurines	Uridylate, orotidylate, 3-phosphoribosyl-uric acid, and xanthine	2.4.2.9
Nicotinate	Nicotinate ribonucleotide	2.4.2.11
Nicotinamide	Nicotinamide ribonucleotide	2.4.2.12
Quinolinate	Quinolinate ribonucleotide + CO_2	—
Imidazoleacetate (+ ATP)	Phosphoribosyl imidazoleacetate (+ ADP + P_i)	—
Histamine (+ ATP)	Phosphoribosyl histamine (+ ADP + P_i)	—
Anthranilate	Phosphoribosyl anthranilate	—
ATP	1-(5'-Phosphoribosyl)-ATP	—

B. MECHANISMS

Adenine, hypoxanthine-guanine, and (under certain conditions) quinolinate phosphoribosyltransferases are all ordered bi-bi reactions whose initial velocity data are apparently compatible only with a ping-pong bi-bi mechanism. Isotope exchange studies of PP-ribose-P glutamine amidotransferase suggest that its mechanism is neither random nor ping pong, but initial velocity and product inhibition data have not been reported. Isotope exchange studies of ATP phosphoribosyltransferase suggest a ping-pong mechanism, but initial velocity data are not consistent with this conclusion; this discrepancy has not been further studied. Further generalizations regarding differences or similarities among this group of reactions cannot yet be made (see reference *11*).

References

1. Meister, A., "Biochemistry of the Amino Acids," 2nd ed., Vols. 1 and 2. Academic Press, New York, 1965.
2. Blakley, R. L., "The Biochemistry of Folic Acid and Related Pteridines. Wiley (Interscience), New York, 1969.
3. Huennekens, F. M., *Biochemistry* **2,** 151 (1964).
4. Meister, A., *Harvey Lect.* **63,** 139 (1969).
5. Mizobuchi, K., and Buchanan, J. M., *J. Biol. Chem.* **243,** 4853 (1968).
6. Mizobuchi, K., Kenyon, G. L., and Buchanan, J. M., *J. Biol. Chem.* **243,** 4863 (1968).
7. Chu, S. Y., and Henderson, J. F., *Can. J. Biochem.* **50,** 490 (1972)
8. Schroeder, D. D., Allison, A. J., and Buchanan, J. M., *J. Biol. Chem.* **244,** 5856 (1969).
9. French, T. C., Dawid, I. B., Day, R. A., and Buchanan, J. M., *J. Biol. Chem.* **238,** 2171 (1963).
10. Dawid, I. B., French, T. C., and Buchanan, J. M., *J. Biol. Chem.* **238,** 2178 (1963).
11. Henderson, J. F., Brox, L. W., Kelley, W. N., Rosenbloom, F. M., and Seegmiller, J. E., *J. Biol. Chem.* **243,** 2514 (1968).

CHAPTER 6

RIBOSE PHOSPHATE SYNTHESIS

I. Introduction

It is as nucleotides, i.e., ribosyl phosphate derivatives, that most of the reactions of purines and pyrimidines take place, and the origin of the ribosyl phosphate moiety must therefore be discussed. The deoxyribosyl phosphate moiety of deoxyribonucleotides is synthesized by way of ribonucleotides, and this process will be discussed in Chapter 16. The pertinent literature through 1962 has been reviewed in the English edition of Hollman's book (1); references given there will not usually be repeated here.

The pathways by which ribose phosphates are formed will be considered in two stages: the initial formation of pentose phosphates in general, and the formation of PP-ribose-P and ribose-1-P.

II. Direct Phosphorylation of Pentoses

Direct phosphorylation of pentoses may lead either to ribose-5-P, or to other pentose 5-phosphates from which it can be derived. Three such reactions may be listed:

$$\text{Ribose} + \text{ATP} \rightarrow \text{ribose-5-P} + \text{ADP}$$

$$\text{Ribulose} + \text{ATP} \rightarrow \text{ribulose-5-P} + \text{ADP}$$

$$\text{Xylulose} + \text{ATP} \rightarrow \text{xylulose-5-P} + \text{ADP}$$

All three enzymes have been reported in various microorganisms, and low activities have also been found in animal tissues such as liver and thyroid. These enzymes may be involved to some extent in the utilization of pentoses in diets or growth media, but there is little evidence that the utilization of nutritional pentoses is of major quantitative importance in the formation of nucleotide pentose phosphate in most organisms.

III. The Pentose Phosphate Pathway

This pathway is sometimes referred to as a "shunt," which tends to imply that it is an alternative to the glycolytic pathway and of secondary importance. The glycolytic and pentose phosphate pathways are each essential, however, and should be considered as separate, though interdependent, pathways.

A. Cyclic or Linear Operation of the Pathway?

The early investigations which resulted in the elucidation of the pentose phosphate pathway have been reviewed by Dickens (2) and will not be discussed here. The pathway and its interconnections with the glycolytic pathway are shown in Fig. 6-1. It is presented here as a cyclic pathway, in which hexose phosphate is converted to pentose phosphates via NADP+-linked dehydrogenases, and in which pentose phosphates can be converted to hexose phosphate and triose phosphate. Horecker (3–5) has proposed an alternative view of the pentose phosphate pathway: that of two separate pathways or branches converging on pentose phosphates. This interpretation is shown in Table 6–I. In the oxidative branch, glucose-6-P is irrever-

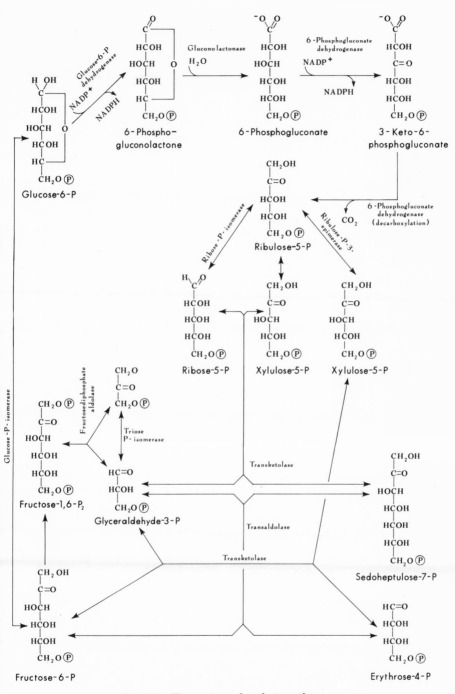

FIG. 6-1. The pentose phosphate pathway.

TABLE 6-I

THE PENTOSE PHOSPHATE PATHWAY

I. The Oxidative Branch

$$\text{Glucose 6-phosphate} + NADP^+ \rightarrow \text{6-phosphogluconolactone} + NADPH + H^+$$
$$\text{6-Phosphogluconolactone} + H_2O \rightarrow \text{6-phosphogluconate}$$
$$\text{6-Phosphogluconate} + NADP^+ \rightarrow \text{ribulose 5-phosphate} + CO_2 + NADPH + H^+$$

Sum: $\text{Glucose 6-phosphate} + 2\ NADP^+ + H_2O \rightarrow \text{ribulose 5-phosphate} + CO_2 + 2\ NADPH + H^+$

II. The Nonoxidative Branch

$$2\ \text{Glucose 6-phosphate} \rightleftharpoons 2\ \text{fructose 6-phosphate}$$

Fructose 6-phosphate + glyceraldehyde 3-phosphate $\rightleftharpoons$ xylulose 5-phosphate + erythrose 4-phosphate
Fructose 6-phosphate + erythrose 4-phosphate $\rightleftharpoons$ sedoheptulose 7-phosphate + glyceraldehyde 3-phosphate
Sedoheptulose 7-phosphate + glyceraldehyde $\rightleftharpoons$ xylulose 5-phosphate + ribose 5-phosphate
3-phosphate

2 Xylulose 5-phosphate $\rightleftharpoons$ 2 ribulose 5-phosphate
Glyceraldehyde 3-phosphate $\rightleftharpoons$ 3 ribose 5-phosphate

Sum: 2 Glucose 6-phosphate + glyceraldehyde 3-phosphate $\rightleftharpoons$ 3 ribose 5-phosphate sibly converted to ribulose-5-P, while in the nonoxidative branch (which is also referred to as the transaldolase–transketolase pathway), hexose phosphate and triose phosphate are converted to pentose phosphates; all of the reactions of the nonoxidative branch are freely reversible.

B. COORDINATION OF TRANSALDOLASE AND TRANSKETOLASE REACTIONS

The metabolism of fructose-6-P by the nonoxidative branch could in theory be initiated by transketolase, with triose phosphate as an acceptor, or by transaldolase, with erythrose-4-P as an acceptor. Dische (6) and Bonsignore et al. (7) found, however, that fructose-6-P was rapidly converted to sedoheptulose-7-P in cell extracts in the absence of added acceptor, and this observation was confirmed with a mixture of purified transaldolase and transketolase; the presence of both enzymes was required (8). It appeared that each enzyme had reacted with fructose-6-P to form the acceptor for the other enzyme, as shown in Fig. 6-2.

Venkataraman and Racker (9, 10) subsequently found that a stable transaldolase–dihydroxyacetone complex was formed as an intermediate in the transaldolase-catalyzed reaction; this was confirmed by Horecker and his co-workers (4, 11) with crystalline transaldolase. Horecker found that the purified complex had a very high affinity for erythrose-4-P; incubation of the complex with $10^{-6} M$ erythrose-4-P, a concentration which could be attained through the action of transketolase on fructose-6-P, resulted in the ready formation of sedoheptulose-7-P, as shown in Fig. 6-3. The glyceraldehyde-3-P could then react with the transketolase–

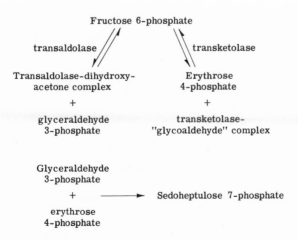

FIG. 6-2. Coordination of the transaldolase and transketolase reactions.

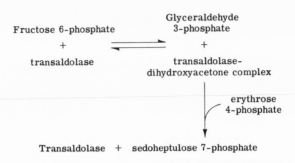

FIG. 6-3. The transaldolase reaction.

"glycoaldehyde" complex, resulting in the formation of xylulose-5-P and the release of free transketolase. [The transketolase–"glycoaldehyde" complex has been shown by Racker (12) and Krampitz et al. (13) to be an enzyme-bound dihydroxyethyl derivative of thiamine pyrophosphate.] The driving force behind the coupled transaldolase–transketolase reaction with fructose-6-P is therefore the great affinity of the transaldolase–dihydroxyacetone complex for erythrose-4-P.

C. Evaluation of Oxidative and Nonoxidative Segments of the Pathway

It is generally agreed that the pentose phosphate pathway is the major route by which the precursors of nucleotide and nucleic acid pentoses are formed, but there is less agreement regarding the relative contributions of the oxidative and nonoxidative branches of the pathway. In a number of investigations of the relative importance of the two branches, specifically labeled glucose was administered, the labeling patterns in nucleic acid pentoses were examined, and, making certain assumptions, the relative contribution of each branch was deduced. In some of these studies the assumption was made that the reactions of both branches of the pathway operate in an essentially irreversible manner; however, Wood et al. (14) suggest that many of the results obtained in such experiments can also be explained on the basis of isotope exchange via the readily reversible transaldolase and transketolase reactions. They also point out the effect of recycling of glucose through pentose phosphates back to hexose phosphates by the pathway; the "recycled" hexose phosphate would be expected to exhibit a different labeling pattern from that of the original hexose. This "recycled" hexose phosphate, upon reentering the pentose phosphate pathway, would then give rise to pentose phosphate exhibiting at least some degree of "randomization" of label.

Katz and Rognstad (15) have more recently examined this problem and have developed a general and relatively simple model for glucose metabolism; using this model and published data for labeling patterns of pentoses formed from glucose-^{14}C, they concluded that the *net flow* of carbon in the nonoxidative branch of the pentose phosphate pathway was from pentose to hexose, and that the *net synthesis* of pentose phosphates from hexose phosphates occurred exclusively by means of the oxidative branch of the pathway.

At the present time, it is rather difficult to evaluate the various conflicting interpretations of the available data. Perhaps the simplest view of the pentose phosphate pathway is that the oxidative branch functions to provide NADPH, a source of hydrogen for reductive biosyntheses, and to provide pentose phosphates; the nonoxidative branch functions to reutilize pentose phosphates, which may be derived from ribonucleoside degradation, or are synthesized oxidatively in excess of those required for nucleotide synthesis; it may also provide pentose phosphates. Further studies are required to settle the many questions that still remain.

That either the oxidative or the nonoxidative branch alone is able to provide sufficient pentose phosphates to support growth has been demonstrated by studies of various microorganisms. Thus, *Candida* (*Torula*) *utilis* apparently uses the oxidative pathway only, even though it contains transaldolase and transketolase (16, 17). In contrast, *Alcaligenes faecalis* and *Pseudomonas saccharophilia* lack the oxidative branch (18, 19). In some cases still different pathways of pentose phosphate synthesis may occur in bacteria. Human erythrocytes deficient in glucose-6-P dehydrogenase can synthesize pentose phosphates adequately via the nonoxidative branch (20).

IV. Ribose 1-Phosphate Synthesis

It will be shown in Chapters 8 and 12 that purines and pyrimidines may be converted to ribonucleosides by reaction with ribose-1-P, and directly to ribonucleotides by reaction with 5-phosphoribosyl 1-pyrophosphate (PP-ribose-P).

Ribose-1-P can be formed from ribose-5-P by a phosphoribomutase reaction:

$$\text{Ribose-5-P} \rightleftharpoons \text{ribose-1-P}$$

This reaction, which can be catalyzed by phosphoglucomutase, has been studied in detail by Klenow and co-workers (21–25). They demonstrated that the reaction involves ribose 1,5-diphosphate as an obligatory intermediate, and a phosphoenzyme:

$$\begin{array}{ccccc}
\text{Ribose-5-P} & & \text{ribose 1,5-diphosphate} & & \text{ribose-1-P} \\
+ & \rightleftharpoons & + & \rightleftharpoons & + \\
\text{phosphoenzyme} & & \text{dephosphoenzyme} & & \text{phosphoenzyme}
\end{array}$$

Klenow concluded that both phosphoglucomutase and phosphoribomutase activities reside in a single protein; the ratio of the two activities remained constant during purification of the enzyme from muscle, both activities were stimulated by Mg^{2+} and cysteine, both exhibited the same pH optimum, and both were inhibited by inorganic anions. On the other hand, Guarino and Sable (26, 27) have purified a specific phosphoribomutase from uterine muscle; this mutase showed no requirement for divalent cations. At the present time, it is difficult to evaluate the relative physiological significance of the two mutases in the interconversion of the ribose phosphates.

Ribose 1-phosphate can also be formed by the phosphorolysis of nucleosides, catalyzed by the nucleoside phosphorylases (see Chapters 8, 10, and 12). The ribose-1-P so derived from ribonucleosides may also be con-

$$\text{Ribosyl purine (pyrimidine)} + P_i \rightleftharpoons \text{ribose-1-P} + \text{purine (pyrimidine)}$$

verted to lactate and phosphorylated glycolytic intermediates (28–31) and to PP-ribose-P (32).

V. The Formation of Phosphoribosyl Pyrophosphate

PP-Ribose-P is the most important ribose phosphate donor for purine metabolism (see Chapters 7, 8) and participates in several important reactions of pyrimidine metabolism (Chapters 11, 12); it also transfers this group to a number of other acceptors (Chapter 5). In the course of studies of purine ribonucleotide biosynthesis, a product of the reaction of ribose-5-P and ATP was isolated and eventually identified as 5-phosphoribosyl 1-pyrophosphate. The pyrophosphate group is in the α-configuration, is quite labile, and almost certainly reacts enzymatically as the magnesium complex. It has been chemically synthesized by Tener and Khorana (33).

A. RIBOSE PHOSPHATE PYROPHOSPHOKINASE

The formation of PP-ribose-P is catalyzed by the enzyme, ribose phosphate pyrophosphokinase:

This enzyme was first studied by Remy et al. (34) and by Kornberg et al. (35) in extracts of pigeon liver, and was shown to require both Mg^{2+} and inorganic phosphate. MgATP is the true substrate of the reaction (36, 37), and a pyrophosphoryl–enzyme complex has been proposed as an intermediate in the reaction (38). Approximate Michaelis constants for the Ehrlich ascites tumor cell enzyme are 3.5 mM for MgATP, and 0.02 to 0.2 mM for ribose-5-P. For the Salmonella enzyme, the constant for ATP ranges from 0.1 to 2.9 mM. The latter enzyme has been shown to catalyze the reverse reaction as well, although the animal enzyme has been considered to be irreversible. The equilibrium constant was about 38, and Michaelis constants for the reverse reaction were 1.5 to 4 $\times$ 10^{-4} M for adenylate and 5 $\times$ 10^{-5} to 1.3 $\times$ 10^{-4} for PP-ribose-P. The Salmonella enzyme has a molecular weight of about 540,000 (37).

Both O-adenylyl methylenediphosphonate and ribose 5-phosphorothioate can serve as substrates for the pyrophosphokinase, and the PP-ribose-P analogues so formed are substrates for several reactions which require PP-ribose-P (39).

B. REGULATION OF RIBOSE PHOSPHATE PYROPHOSPHOKINASE

Studies of this enzyme from Ehrlich ascites tumor cells, erythrocytes, and Salmonella typhimurium have indicated a number of features which are at least potential mechanisms for the control of its activity. Thus, although Mg^{2+} is needed to form the actual substrate, MgATP, free Mg^{2+} also stimulates enzyme activity (36, 37). One of the most interesting effectors is inorganic phosphate, for which there is an absolute requirement. With the tumor cell enzyme, 50 mM phosphate is required for maximal activity with an apparent Michaelis constant of 3.3 mM (36, 40). A biphasic response of the bacterial enzyme to phosphate was observed with apparent Michaelis constants of 2.3 and 40 mM (37). In both cases, arsenate was a weak substitute. Wong and Murray (40) found that the Michaelis constant for ATP did not vary with phosphate concentration, whereas that for ribose-5-P did vary. Hershko et al. (41) and others have also shown stimulation of enzyme activity by phosphate in intact erythrocytes and Ehrlich ascites tumor cells incubated in media containing varying concentrations of phosphate.

Ribose phosphate pyrophosphokinase from tumor cells is inhibited by excess ATP, ADP, CTP, and other triphosphates, all of which are competitive with respect to MgATP; inhibition by MgGTP is not competitive (40). The erythrocyte enzyme is inhibited by ADP, and this is relieved by increased phosphate; increased ATP has no effect. GDP also inhibits, but this is not affected by phosphate; 2,3-diphosphoglycerate is also a potent inhibitor (*20*). Hershko *et al.* (*20*) have also found that the maximum rate of ribose phosphate pyrophosphokinase activity in erythrocyte extracts is 200 times the apparent rate of this enzyme in intact erythrocytes. They believe that the enzyme is normally under very strong inhibition by nucleotides and other metabolites, and that phosphate counteracts these inhibitory effects. Atkinson and co-workers (*42, 43*) found that a bacterial ribose phosphate pyrophosphokinase is controlled by the relative amounts of ATP, ADP, and adenylate (see Chapter 3, Section VII), and by nonpurine end products of PP-ribose-P metabolism, such as histidine and tryptophan (see Chapter 5). This has also been reported by Switzer (*44*).

C. RIBOSE 5-PHOSPHATE AVAILABILITY

The synthesis of PP-ribose-P in Ehrlich ascites tumor cells incubated *in vitro* is a function of the initial extracellular glucose concentration up to a maximum of 0.55 mM, and its initial rate of synthesis is about 3% of the initial rate of glycolysis. The first limiting factor in the synthesis of ribose-5-P from glucose is hexokinase, the apparent Michaelis constant of which is approximately $10^{-4}\ M$. The oxidative branch of the pentose phosphate pathway is saturated at an extracellular glucose concentration of $2.5 \times 10^{-5}\ M$ (*32*). Erythrocytes deficient in glucose-6-P dehydrogenase have no deficiency of PP-ribose-P, however (*20*). Both tumor cells and erythrocytes show accelerated PP-ribose-P synthesis in the presence of methylene blue; as a consequence of accelerated NADPH reoxidation, ribose-5-P synthesis via oxidation of glucose-6-P takes place at an increased rate. In erythrocytes, this is observed only when the rate of ribose phosphate pyrophosphokinase is increased by incubation in high phosphate. Glycolytic inhibitors and anaerobiosis inhibit PP-ribose-P synthesis from glucose and from ribonucleosides in tumor cells (*31, 32, 45*), but it is not known whether these treatments affect ribose-5-P concentrations or those of various effectors.

D. INHIBITION OF PHOSPHORIBOSYL PYROPHOSPHATE SYNTHESIS BY DRUGS

The formation of PP-ribose-P is inhibited by the triphosphates of a variety of adenosine analogues, and this inhibition has been proposed as a possible biochemical explanation for the antitumor activity of some of these analogues. Among the compounds which have been shown to inhibit the formation of PP-ribose-P are xylosyladenine, 3′-deoxyadenosine (cordycepin), tubercidin, and formycin. Psicofuranine and decoyinine, two adenosine analogues which are not converted to the corresponding triphos-

TABLE 6–II

ENZYMES OF RIBOSE PHOSPHATE SYNTHESIS

Enzyme Commission number	Systematic name	Trivial name
1.1.1.49	D-Glucose-6-phosphate: NADP oxidoreductase	Glucose-6-phosphate dehydrogenase
3.1.1.17	D-Glucono-δ-lactone hydrolase	Gluconolactonase
1.1.1.44	6-Phospho-D-gluconate: NAD: oxidoreductase (decarboxylating)	Phosphogluconate dehydrogenase (decarboxylating)
1.1.1.43	6-Phospho-D-gluconate: NAD (P) oxidoreductase	Phosphogluconate dehydrogenase
5.1.3.1	D-Ribulose-5-phosphate 3-epimerase	Ribulosephosphate 3-epimerase
5.3.1.6	D-Ribose-5-phosphate ketol-isomerase	Ribosephosphate isomerase
2.2.1.1.	Sedoheptulose-7-phosphate: D-glyceraldehyde-3-phosphate glycolaldehydetransferase	Transketolase, glycolaldehyde-transferase
2.2.1.2.	Sedoheptulose-7-phosphate: D-glyceraldehyde-3-phosphate dihydroxyacetonetransferase	Transaldolase, dihydroxyacetonetransferase
5.3.1.1.	D-Glyceraldehyde-3-phosphate ketol-isomerase	Triosephosphate isomerase
4.1.2.13	Fructose-1,6-diphosphate: D-glyceraldehyde-3-phosphatelyase	Fructosediphosphate aldolase
2.7.1.11	ATP: D-fructose-6-phosphate 1-phosphotransferase	Phosphofructokinase
3.1.3.11	D-Fructose-1,6-diphosphate 1-phosphohydrolase	Hexosediphosphatase
5.3.1.9	D-Glucose-6-phosphate ketol-isomerase	Glucosephosphate isomerase
2.7.5.1	α-D-Glucose-1,6-diphosphate: α-D-glucose-1-phosphate phosphotransferase	Phosphoglucomutase
	—	Phosphoribomutase
2.7.6.1	ATP: D-ribose-5-phosphate pyrophosphotransferase	Ribosephosphate pyrophosphokinase
2.7.1.1	ATP: D-hexose-6-phosphotransferase	Hexokinase

phates, also inhibited the formation of PP-ribose-P by extracts of *Streptococcus faecalis*. This subject is reviewed by Balis (*46*). The structures of some of these compounds are given in Chapter 17, together with further discussion of their effects and metabolism.

The enzymes of ribose phosphate synthesis are listed in Table 6–II.

References

1. Hollmann, S., "Non-Glycolytic Pathways of Metabolism of Glucose." Academic Press, New York, 1964.
2. Dickens, F., *Ann. N.Y. Acad. Sci.* **75**, 71 (1958).
3. Horecker, B. L., and Mehler, A. H., *Annu. Rev. Biochem.* **24**, 207 (1955).
4. Horecker, B. L., *Harvey Lect.* **57**, 35 (1962).
5. Horecker, B. L., *J. Chem. Educ.* **24**, 244 (1965).
6. Dische, Z., *Ann. N.Y. Acad. Sci.* **75**, 129 (1958).
7. Bonsignore, A., Pontremoli, S., Fornaini, G., and Grazi, E., *Ital. J. Biochem.* **6**, 241 (1957).
8. Pontremoli, S., Bonsignore, A., Grazi, E., and Horecker, B. L., *J. Biol. Chem.* **235**, 1881 (1960).
9. Venkataraman, R., and Racker, E., *J. Biol. Chem.* **236**, 1876 (1961).
10. Venkataraman, R., and Racker, E., *J. Biol. Chem.* **236**, 1883 (1961).
11. Horecker, B. L., Pontremoli, S., Ricci, C., and Cheng, T., *Proc. Nat. Acad. Sci. U.S.* **47**, 1949 (1961).
12. Racker, E., *in* "The Enzymes" (P. D. Boyer, H. Lardy, and K. Myrbäck, eds.), 2nd rev. ed., Vol. 5, p. 397. Academic Press, New York, 1961.
13. Krampitz, L. O., Suzuki, I., and Greull, G., *Fed. Proc., Fed. Amer. Soc. Exp. Biol.* **20**, 971 (1961).
14. Wood, H. G., Katz, J., and Landau, B. R., *Biochem. Z.* **338**, 809 (1963).
15. Katz, J., and Rognstad, R., *Biochemistry* **6**, 2227 (1967).
16. Sowden, J. C., Frankel, S., Moore, B. H., and McClary, J. E., *J. Biol. Chem.* **206**, 547 (1954).
17. Barker, G. R., and Nicholson, B. H., *Biochem. J.* **91**, 326 (1964).
18. Brenneman, F. N., Vishniac, W., and Volk, W. A., *J. Biol. Chem.* **235**, 3357 (1960).
19. Fossitt, D. P., and Bernstein, I. A., *J. Bacteriol.* **86**, 1326 (1963).
20. Hershko, A., Razin, A., and Mager, J., *Biochim. Biophys. Acta* **184**, 64 (1969).
21. Klenow, H., and Larsen, B., *Arch. Biochem. Biophys.* **37**, 488 (1952).
22. Klenow, H., *Arch. Biochem. Biophys.* **46**, 186 (1953).
23. Klenow, H., *Arch. Biochem. Biophys.* **58**, 288 (1955).
24. Klenow, H., and Emberland, R., *Arch. Biochem. Biophys.* **58**, 276 (1955).
25. Klenow, H., Emberland, R., and Plesner, P., *Acta Chem. Scand.* **8**, 1103 (1954).
26. Guarino, A. J., and Sable, H. Z., *J. Biol. Chem.* **215**, 515 (1955).
27. Guarino, A. J., and Sable, H. Z., *Biochim. Biophys. Acta* **20**, 201 (1956).
28. Lionetti, F. J., and Fortier, N. L., *Arch. Biochem. Biophys.* **103**, 15 (1963).
29. Shafer, A. W., and Bartlett, G. R., *J. Clin. Invest.* **41**, 690 (1962).
30. Bishop, C., *J. Biol. Chem.* **239**, 1053 (1964).
31. Fontenelle, L. J., and Henderson, J. F., *Can. J. Biochem.* **47**, 419 (1969).
32. Henderson, J. F., and Khoo, M. K. Y., *J. Biol. Chem.* **240**, 2349 (1965).

33. Tener, G. M., and Khorana, H. G., *J. Amer. Chem. Soc.* **80**, 1999 (1958).
34. Remy, C. N., Remy, W. T., and Buchanan, J. M., *J. Biol. Chem.* **217**, 885 (1955).
35. Kornberg, A., Lieberman, I., and Simms, E. S., *J. Biol. Chem.* **215**, 389 (1955).
36. Murray, A. W., and Wong, P. C. L., *Biochem. Biophys. Res. Commun.* **29**, 582 (1967).
37. Switzer, R. L., *J. Biol. Chem.* **244**, 2854 (1969).
38. Switzer, R. L., *Biochem. Biophys. Res. Commun.* **32**, 320 (1968).
39. Murray, A. W., Wong, P. C. L., and Friedricks, B., *Biochem. J.* **112**, 741 (1969).
40. Wong, P. C. L., and Murray, A. W., *Biochemistry* **8**, 1608 (1969).
41. Hershko, A., Razin, A., Shoshani, T., and Mager, J., *Biochim. Biophys. Acta* **149**, 59 (1967).
42. Atkinson, D. E., and Fall, L., *J. Biol. Chem.* **242**, 3241 (1967).
43. Klungsoyr, L., Hagemen, J. H., Fall, L., and Atkinson, D. E., *Biochemistry* **7**, 4035 (1968).
44. Switzer, R. L., *Fed. Proc., Fed. Amer. Soc. Exp. Biol.* **26**, 560 (1967).
45. Henderson, J. F., and Khoo, M. K. Y., *J. Biol. Chem.* **240**, 2363 (1965).
46. Balis, M. E., "Antagonists and Nucleic Acids." North-Holland Publ., Amsterdam, 1968.

PART II

PURINE RIBONUCLEOTIDE METABOLISM

Purine ribonucleotides may be synthesized *de novo* from amino acids and other small molecules (Chapter 7), or formed from preformed bases and ribonucleosides which may be derived from the diet or found in the environment of cells (Chapter 8). Ribonucleotides of adenine and guanine may be converted one into the other (Chapter 9). Uric acid is formed from purine ribonucleotides by dephosphorylation, deamination, cleavage of glycosidic bonds, and oxidation, and in turn is converted to CO_2 and NH_3 in some organisms (Chapter 10).

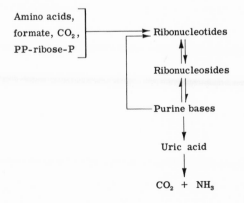

CHAPTER 7

PURINE BIOSYNTHESIS *DE NOVO*

I. Introduction and Early History

The synthesis of purines from nonpurine precursors ("purine biosynthesis *de novo*") is one of the oldest areas of biochemistry. Evidence for the existence of this process became available about 150 years ago, and it has been explicitly studied for some 80 to 90 years. It was not until about 1960, however, that the enzymatic reactions of purine biosynthesis *de novo* were identified, and this pathway is still an area of current research. Some of the very early work which led to the recognition of the existence of purine biosynthesis *de novo* will be reviewed briefly; McCrudden (*1*) presents a fuller account of these developments.

Uric acid (2,6,8-trioxypurine) was discovered in 1776 in human urine and bladder stones by Scheele (*2*) and by Bergmann, and by 1805 it had been found in the excreta of a wide variety of animals. Further studies during the first half of the nineteenth century showed that uric acid was the chief end product of nitrogen metabolism in birds and some reptiles, and this work implied that reactions for its biosynthesis *de novo* existed. Despite great interest in uric acid during this period, its structure was not definitively established until 1882, when it was synthesized by Fischer; its elementary formula had been determined by Liebig in 1834, however, and Medicus had proposed the correct structure in 1875.

Nucleic acids were discovered in 1868–1869, and the biosynthesis of nucleic acid purines *de novo* was explicitly proposed by Miescher in 1874. He had observed that Rhine salmon converted much of their tissue mass (i.e., protein) into the nucleic acids of spermatozoa during migration upstream, although they do not eat during this period. In 1885 a tenfold increase in the total purine content of silkworm eggs was observed to occur during development, and similar observations were later made with eggs of other insects and of birds.

The existence of a purine biosynthetic pathway in mammals was soon established through the use of purine-free diets. In 1891 mice were first kept on such diets for extended periods with no ill-effects, and similar experiments were subsequently performed with other animals and man.

These early studies also contributed to an understanding of the site of uric acid biosynthesis and led to several hypothetical reaction schemes. In 1877 several investigators showed that although the feeding of urea, amino acids, or ammonium salts to mammals led to an increase in the excretion of urea, the feeding of these compounds to birds resulted instead in an increased excretion of uric acid. Minkowski showed in 1886 that hepatectomized birds did not synthesize uric acid from such precursors, and because urinary lactate was elevated in such birds, he suggested that uric acid was synthesized from lactate and ammonia.

A more widely accepted scheme for uric acid synthesis was proposed in 1902 by Wiener. In this hypothesis urea was condensed with hydroxy-malonic acid to form dialuric acid, which was then condensed with a second molecule of urea.

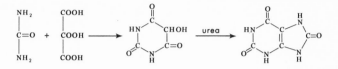

Wiener's scheme was accepted for many years, although some later workers disputed the data on which it was based. Others suggested instead that histidine, arginine, or pyrimidines were purine precursors. Although attempts were made between 1900 and 1940 to obtain more information concerning the identity of the precursors of uric acid by means of feeding experiments, this field did not advance very much until the advent of isotopically labeled compounds.

A. PRECURSORS OF THE PURINE RING

Between 1943 and 1956 the biochemical source of each of the atoms of the purine ring was established, and between 1950 and 1960 most of the reactions of the purine biosynthetic pathway were elucidated. The following discussion of purine biosynthesis *de novo* is based largely on the studies of G. R. Greenberg, J. M. Buchanan, and their associates. The pigeon and chicken liver systems which they employed are still the most studied, although the intermediates and enzymes of this pathway have also been recognized in other organisms as well. A number of reviews have been published in which the work of this period is discussed, and references contained in them will in general not be repeated here (*3-7*).

Early studies with labeled compounds were conducted simply to determine if administered isotopes could be found in urinary uric acid or in the purines of tissue nucleic acids. More information was to be gained, however, by localizing the label in specific positions in the purine ring, and methods were therefore developed by which the uric acid molecule could be chemically degraded and each ring atom separately isolated.

The origins of the ring nitrogens of uric acid were first studied in 1943 by Barnes and Schoenheimer. Little ^{15}N from urea was incorporated into visceral nucleic acid purines of the rat, and urea could be eliminated as a potential precursor, thereby disproving Weiner's hypothesis. ^{15}N-Ammonium citrate was a good precursor of purines, however, and because the specific activity of purines was considerably greater than that of histidine and arginine, these amino acids were eliminated as possible precursors. This point was later confirmed directly by the use of ^{15}N-histidine. A key precursor was discovered in 1947 when Shemin and Rittenberg found that isotope from ^{15}N-glycine predominantly labeled position 7 of urinary uric acid.

Studies of the sources of the ring carbon atoms of uric acid were begun in 1948 by Sonne, Buchanan, and Delluva, who administered a variety of ^{13}C-labeled compounds to pigeons, isolated uric acid from the excreta, and degraded it to identify the position of the label. They showed that (a) CO_2 and HCO_3^- were precursors of C-6 of uric acid; (b) formate labeled C-2 and C-8 equally; and (c) the carboxyl carbon of glycine was a precursor of C-4. This work was later confirmed and extended by Karlson and Barker, who found also that the methylene carbon of glycine was a precursor of C-5. Glycine, therefore, was incorporated intact into the C-4, C-5, and N-7 positions of uric acid.

Because of rapid distribution throughout the amino acid pool of ^{15}N from labeled ammonium salts and individual amino acids, the specific sources of nitrogens other than N-7 could not be determined in whole animals. Following the development of a cell-free system in which purine biosynthesis *de novo* could be achieved (see below), the rate of this process was shown to be markedly stimulated by supplementation with aspartate and glutamine. Experiments with these amino acids labeled with ^{15}N indicated that the former contributed N-1, while the amide group of glutamine contributed N-3 and N-9.

The overall result of these precursor studies may be summarized as follows:

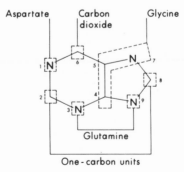

One-carbon units

B. DEVELOPMENT OF A CELL-FREE SYSTEM

Detailed studies of the reactions of purine biosynthesis *de novo* and the isolation and identification of the intermediates of this pathway required the use of an *in vitro*, and preferably cell-free, system. Shortly after new techniques made these studies feasible, Greenberg developed such a system, based on earlier studies by H. A. Krebs. The latter had turned briefly to the study of uric acid synthesis in birds following his elucidation of the urea cycle in mammals in 1932. He found that although uric acid synthesis could be demonstrated in liver slices of some birds, pigeon liver formed hypoxanthine instead. (It is converted to uric acid in the kidneys of these birds.) Hypoxanthine synthesis in pigeon liver slices was stimulated by the addition of glutamine, but no synthesis was detected in broken-cell preparations. This work was particularly important because hypoxanthine is two steps closer to the origin of the purine biosynthetic pathway than is uric acid, and led to the choice of pigeon liver for subsequent studies.

By addition of the then known precursors of uric acid plus amino acids and energy and ribose sources, Greenberg was able to achieve the synthesis of hypoxanthine in homogenates of pigeon liver. He later used the supernatant from high-speed centrifugation of the homogenate, and Buchanan extended this work by precipitation of the purine-synthesizing enzymes with ethanol at low temperature.

C. DISCOVERY OF KEY COMPOUNDS

The elucidation of the sources of the uric acid ring atoms and development of a soluble system capable of purine biosynthesis *de novo* prepared

TABLE 7–I

NAMES OF THE INTERMEDIATES OF PURINE BIOSYNTHESIS *de Novo*

Old	Enzyme Commission
Phosphoribosylamine (PRA)	Ribosylamine phosphate
Glycineamide ribonucleotide (GAR)	Phosphoribosyl glycineamide
Formylglycineamide ribonucleotide (FGAR)	Phosphoribosyl formylglycineamide
Formylglycineamidine ribonucleotide (FGAM)	Phosphoribosyl formylglycineamidine
Aminoimidazole ribonucleotide (AIR)	Phosphoribosyl aminoimidazole
Aminoimidazolecarboxylate ribonucleotide (CAIR)	Phosphoribosyl aminoimidazole carboxylate
Aminoimidazole succinocarboxamide ribonucleotide (SAICAR)	Phosphoribosyl aminoimidazole succinocarboxamide
Aminoimidazolecarboxamide ribonucleotide (AICAR)	Phosphoribosyl aminoimidazole carboxamide
Formamidoimidazolecarboxamide ribonucleotide (FAICAR)	Phosphoribosyl formamidoimidazole carboxamide

the way for studies which showed how this ring system was built up. Several early findings were of special importance in this work, in particular the discovery of the true end product of this pathway and the identification of three key intermediates. The naming of these intermediates poses a problem. Established names and abbreviations are found in the literature, but these are not consistent with today's chemical nomenclature conventions. We have chosen to use a terminology based on Enzyme Commission usage, and to use no special abbreviations. The new and old terminology is shown in Table 7–I.

1. Inosinate: The True End Product

The above-mentioned studies of Krebs had shown that hypoxanthine was closer to the purine biosynthetic pathway than was uric acid. In 1951 Greenberg (8) isolated radioactive inosinate, as well as radioactive hypoxanthine, following incubation of ^{14}C-formate in a cell-free system with all of the other necessary precursors (Fig. 7-1). The specific activity of inosinate was higher than that of hypoxanthine during short incubations and approached that of the precursor formate. This work identified the end product as inosinate rather than hypoxanthine, adenylate, or other purine derivative, and thereby showed that ribose phosphate had been added at some point prior to the completion of the biosynthetic pathway.

2. Phosphoribosyl Aminoimidazole Carboxamide

The heterocyclic base, 5-amino-4-imidazole carboxamide, was isolated in 1945 from sulfonamide-inhibited cultures of *Escherichia coli*, but its possible role as a precursor of purines was suggested only somewhat later. This re-

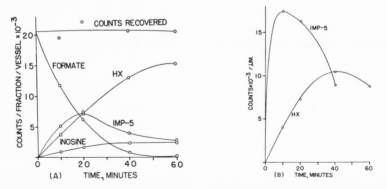

FIG. 7-1. Inosinate, the true end product of purine biosynthesis *de novo* in pigeon liver extracts incubated with substrates and ^{14}C-formate. (A) Total radioactivity of products formed; (B) specific activity of products formed. HX = hypoxanthine; IMP-5 = inosinate. From (8). Reproduced with permission.

lationship was at first questioned because aminoimidazole carboxamide was not utilized by a number of microbial species, but in 1950 it was shown to be incorporated into nucleic acid purines in rats and converted to uric acid by pigeons.

It was subsequently shown that radioactive aminoimidazole carboxamide was converted to inosinate by pigeon liver extracts, and it seemed logical to assume that a ribose phosphate derivative of aminoimidazole carboxamide must be an intermediate. The "ribonucleotide," phosphoribosyl aminoimidazole carboxamide, was synthesized enzymatically and was indeed converted to inosinate by pigeon liver fractions.

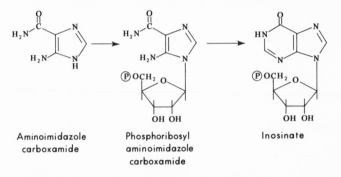

Aminoimidazole carboxamide	Phosphoribosyl aminoimidazole carboxamide	Inosinate

These results indicated not only that phosphoribosyl aminoimidazole carboxamide was a precursor of purine nucleotides, but also that the imidazole portion of the purine was formed prior to the pyrimidine moiety, and that this nonpurine intermediate was involved in the form of a ribose phosphate derivative.

3. Phosphoribosyl Glycineamide and Phosphoribosyl Formylglycineamide

Goldthwait and Greenberg were able in 1954–1956 to identify two intermediates near the beginning of the purine biosynthetic pathway. Incubation of an extract of pigeon liver acetone powder with ribose-5-P, ATP, glutamine, ^{14}C-glycine, and formate led to the accumulation of two new radioactive compounds, A and B. When ^{14}C-formate was used with nonradioactive glycine, only compound B was labeled. These intermediates were identified as phosphoribosyl glycineamide and phosphoribosyl formylglycineamide, respectively.

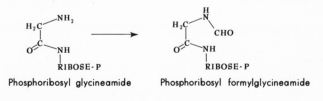

Phosphoribosyl glycineamide   Phosphoribosyl formylglycineamide

The identification of these compounds established the probable phosphoribosyl or "ribonucleotide" nature of all of the intermediates of the purine biosynthetic pathway and indicated that the glycine molecule was the backbone onto which the two rings were built.

The major features of the pathway of purine biosynthesis *de novo* became clear through the discoveries just outlined.

(a) Ribose-5-P + ATP + glutamine + glycine → phosphoribosyl glycineamide, containing C-4, C-5, N-7, N-9

(b) Phosphoribosyl glycineamide + formate → phosphoribosyl formylglycineamide, adding C-8

(c) Phosphoribosyl formylglycineamide + glutamine + CO₂ + aspartate → phosphoribosyl aminoimidazole carboxamide, adding N-3, C-6, and N-1

(d) Phosphoribosyl aminoimidazole carboxamide + formate → inosinate, adding C-2

II. The Reactions of Purine Biosynthesis *de Novo*

A. INTRODUCTION

Successive fractionations of liver extracts have led to the isolation and identification of all of the individual intermediates of the pathway and of the enzymes which synthesize them. These reactions will next be considered in order of their occurrence in this sequence, together with a brief summary of the studies which led to their elucidation.

B. THE REACTIONS

1. Ribosylamine Phosphate Synthesis

Although attempts actually to isolate from biological systems an intermediate between PP-ribose-P and phosphoribosyl glycineamide were unsuccessful, it was found that crude enzyme preparations would convert PP-ribose-P plus glutamine to glutamate and pyrophosphate. It was deduced, therefore, that 1-ribosylamine 5-phosphate was the third product, although it could not be isolated. In fact, enzymatically synthesized ribosylamine phosphate has never been isolated, although the chemically synthesized compound does act as a substrate for the subsequent step in the pathway. Because inosinate is a β-glycoside, this reaction is presumed to involve inversion of the group at C-1 of ribose in the conversion of α-PP-ribose-P to β-ribosylamine phosphate. PP-Ribose-P amidotransferase has been studied extensively, as it occupies an important position as the first enzyme in the pathway of purine biosynthesis *de novo*. It has a molecular weight of 200,000 (*9, 10*), but readily dissociates into two 100,000 molecular weight subunits. Aggregation to the larger unit is promoted by PP-ribose-P. Treatment with sulfhydryl reagents causes dissociation into four subunits of

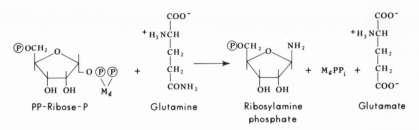

| PP-Ribose-P | Glutamine | Ribosylamine phosphate | Glutamate |

50,000 molecular weight. The 200,000 molecular weight unit contains 10 to 12 atoms of nonheme iron, and the iron has both catalytic and structural functions which are as yet undefined (*10, 11*).

Glutamine, ammonia, and a number of simple alcohols and amines can serve as substrates. The reaction is inhibited by reagents which react with sulfhydryl groups or which chelate iron, and by the glutamine analogue diazo-oxo-norleucine. Such inhibition is retarded in the presence of substrates. The affinity for glutamine is fairly low, the Michaelis constant being about 10^{-3} *M*. That for PP-ribose-P varies between 2×10^{-5} *M* and 2×10^{-4} *M* for different preparations (*9, 12*).

The possible function of this enzyme in the regulation of the pathway is discussed below. It is listed with glutamine amide transfer reactions in Table 5–III and with other phosphoribosyltransferases in Table 5–IV.

The possible physiological role of ammonia as a nitrogen donor for ribosylamine phosphate synthesis has been studied and debated. Although ammonia can react nonenzymatically with PP-ribose-P to form ribosylamine phosphate (*13*), it can also react enzymatically to form this product in cell extracts (*14, 15*) and in intact cells (*16, 17*). Although most work supports the role of PP-ribose-P amidotransferase as the enzyme involved, Reem (*18*) has put forward evidence that a separate enzyme might use ammonia exclusively.

2. Phosphoribosyl Glycineamide Synthesis

The reaction by which phosphoribosyl glycineamide is synthesized is

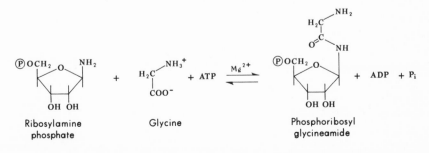

| Ribosylamine phosphate | Glycine | Phosphoribosyl glycineamide |

readily reversible and requires all three components not only for product formation, but also for isotope exchange. In this respect phosphoribosyl glycineamide synthetase resembles several other amide-synthesizing reactions, whose mechanisms have been discussed above (Chapter 5).

This enzyme is relatively small, having a molecular weight of 48,000. The Michaelis constants are 2×10^{-4} M for glycine, 5.6×10^{-5} M for ATP, and about 8×10^{-6} M for ribosylamine phosphate. These concentrations are probably easily attainable in cells (*13*).

3. Phosphoribosyl Formylglycineamide Synthesis

Purine biosynthesis *de novo* was one of the first areas of metabolism in which a folic acid derivative was specifically identified as a cofactor in an enzymatic reaction. The ability of pigeon liver extracts to add formate to phosphoribosyl glycineamide was impaired by treatment with charcoal, but was restored by addition of H_4-folate. Although the complicated interconversions of the H_4-folate coenzymes (see Chapter 5) caused confusion for some time, the specific one-carbon donor for this reaction was eventually identified as 5,10-methenyl H_4-folate. The phosphoribosyl glycineamide formyltransferase reaction itself is irreversible.

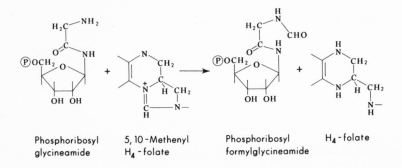

| Phosphoribosyl glycineamide | 5,10-Methenyl H_4-folate | Phosphoribosyl formylglycineamide | H_4-folate |

Recent studies using *Salmonella typhimurium* suggest either that its formyltransferase is atypical, or that it is not an obligate enzyme for purine biosynthesis in this organism. Mutants lacking this enzyme could not be found, nor could a folate coenzyme requirement for phosphoribosyl formylglycineamide synthesis be demonstrated (*19*).

4. Metabolism of Phosphoribosyl Formylglycineamide

The reactions of phosphoribosyl formylglycineamide were considered in detail by Levenberg and Buchanan. A crude pigeon liver extract would convert phosphoribosyl formylglycineamide to inosinate in the presence of other substrates:

Phosphoribosyl   → inosinate
formylglycineamide
+ aspartate + CO_2
+ glutamine
+ ATP + formate

When formate was omitted from this system, a compound accumulated which gave a Bratton-Marshall color reaction identifying it as an aromatic amine. As predicted, this compound was phosphoribosyl aminoimidazole carboxamide.

Phosphoribosyl   →   phosphoribosyl
formylglycineamide   aminoimidazole
+ CO_2 + glutamine   carboxamide
+ aspartate + ATP

When CO_2 and aspartate were omitted from the system in addition to formate, a second aromatic amine accumulated, as shown by a Bratton-Marshall test color different from that given with phosphoribosyl aminoimidazole carboxamide. This amine was identified as phosphoribosyl aminoimidazole. These reaction sequences were then dissected to reveal their component parts.

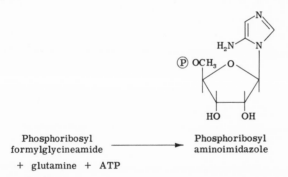

Phosphoribosyl   Phosphoribosyl
formylglycineamide ⟶ aminoimidazole
+ glutamine + ATP

5. Phosphoribosyl Formylglycineamidine Synthesis

Phosphoribosyl formylglycineamidine synthetase catalyzes the irreversible transfer of the amide group of glutamine to phosphoribosyl formylglycineamide to form the corresponding amidine. The molecular weight of the *Salmonella typhimurium* and chicken liver enzymes is 135,000 (*20, 21*). Glutamine binds to the free enzyme with release of the amide nitrogen and probable formation of a γ-glutamyl complex (*22*). A complex with MgATP and phosphoribosyl formylglycineamide can also be formed, and Mizobuchi *et al.* (*23*) concluded that a covalent bond is formed between the enzyme and phosphate derived from ATP. More recently it has been sug-

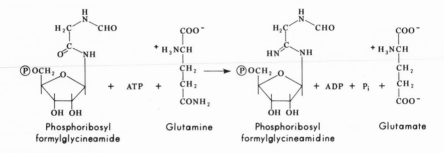

Phosphoribosyl formylglycineamide	Glutamine	Phosphoribosyl formylglycineamidine	Glutamate

gested that the chicken liver enzyme can use ammonia as an alternative substrate, and that it binds at a different site than that to which glutamine binds (*24*). The *Salmonella* enzyme, in contrast, does not appear to use ammonia well (*25*), nor is this reaction supported by ammonia in intact tumor cells (*17*).

6. Phosphoribosyl Aminoimidazole Synthesis

The cyclization of phosphoribosyl formylglycineamidine requires, in addition to ATP, both Mg^{2+} and K^+. The reaction is irreversible. This reac-

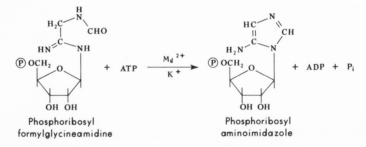

Phosphoribosyl formylglycineamidine	Phosphoribosyl aminoimidazole

tion is catalyzed by phosphoribosyl aminoimidazole synthetase.

7. Phosphoribosyl Aminoimidazole Carboxylate Synthesis

The fixation of CO_2 or HCO_3^- into phosphoribosyl aminoimidazole led to the formation of a new intermediate; because CO_2 was known to furnish the C-6 position, it was identified as phosphoribosyl aminoimidazole carboxylate. The reaction is readily reversible.

For several years there was some confusion concerning the possible role of biotin in this reaction. Phosphoribosyl aminoimidazole accumulated in biotin-deficient yeast, and because this vitamin is known to be involved in several other CO_2 fixation reactions, the obvious inference was made. How-

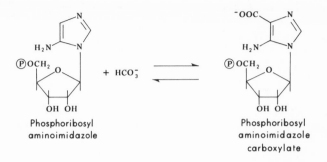

Phosphoribosyl
aminoimidazole

Phosphoribosyl
aminoimidazole
carboxylate

ever, a biotin requirement for the phosphoribosyl aminoimidazole carboxyl-ase reaction could not be shown in rats or pigeons, and phosphoribosyl aminoimidazole did not accumulate in biotin-deficient yeast when aspartate was added. It has now been shown that the biotin is not involved directly (*26*), but is required for aspartate synthesis by the following reactions (see reference *27*):

$$\text{Pyruvate} + CO_2 \xrightarrow{\text{biotin}} \text{oxaloacetate}$$

$$\text{Oxaloacetate} \xrightarrow[\text{amination}]{\text{trans-}} \text{aspartate}$$

Aspartate is required for the subsequent metabolism of phosphoribosyl aminomidazole carboxylate, and phosphoribosyl aminoimidazole accumu-lates because of the ready reversibility of the carboxylation reaction.

8. *Phosphoribosyl Aminoimidazole Succinocarboxamide Synthesis*

The reaction of aspartate with phosphoribosyl aminoimidazole carboxyl-ate produced a compound whose acidity indicated that the dicarboxylic

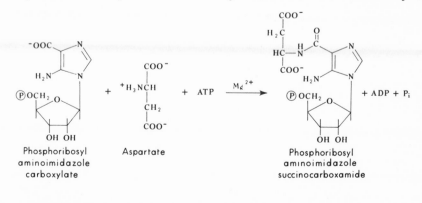

Phosphoribosyl
aminoimidazole
carboxylate

Aspartate

Phosphoribosyl
aminoimidazole
succinocarboxamide

acid groups were uncombined. Titration data revealed that an amide bond had formed between the amino group of aspartate and the carboxyl group of phosphoribosyl aminoimidazole carboxylate forming phosphoribosyl aminoimidazole succinocarboxamide. This enzyme has recently been studied by Fisher (*28*).

9. Phosphoribosyl Aminoimidazole Carboxamide Synthesis

Phosphoribosyl aminoimidazole succinocarboxamide is cleaved in a reversible manner to fumarate and the well-characterized intermediate, phosphoribosyl aminoimidazole carboxamide. The enzyme catalyzing this

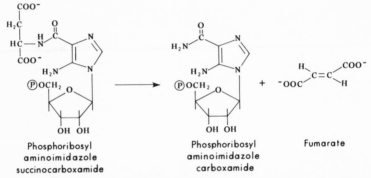

Phosphoribosyl   Phosphoribosyl   Fumarate
aminoimidazole   aminoimidazole
succinocarboxamide   carboxamide

reaction is called adenylosuccinate lyase because a variety of evidence, to be discussed in Chapter 9, indicates that it is identical with that which cleaves adenylosuccinate, an obligatory intermediate in the conversion of inosinate to adenylate. It is most unusual for a single enzyme to participate in two pathways, (or two parts of the same pathway) in this manner.

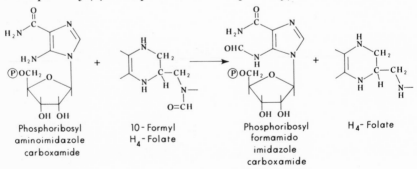

Phosphoribosyl   10-Formyl   Phosphoribosyl H_4-Folate
aminoimidazole H_4-Folate   formamido
carboxamide   imidazole
                           carboxamide

10. Phosphoribosyl Formamidoimidazole Carboxamide Synthesis

Phosphoribosyl aminoimidazole carboxamide formyltransferase catalyzes the addition of a second formate group; it becomes the C-2 position of the purine ring.

11. Inosinate Synthesis

The final step in purine biosynthesis *de novo* is the cyclization of phosphoribosyl formamido imidazole carboxamide to form inosinate. Until

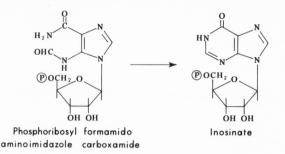

Phosphoribosyl formamido   Inosinate
aminoimidazole carboxamide

recently it was thought likely that this and the preceding reaction were catalyzed by the same enzyme, or that the cyclization was nonenzymatic. However, Gots *et al.* (*29*) have isolated separate bacterial mutants for each. The two enzymes are closely associated and may be part of a single protein aggregate.

III. Summary of the Pathway

Figure 7-2 summarizes this discussion of the reactions of purine biosynthesis *de novo*.

A. UNIVERSALITY

This discussion of the reactions of purine biosynthesis *de novo* is based mainly on work with pigeon liver systems. However, the universality in nature of this scheme of purine synthesis seems fairly certain. Wherever individual reactions or intermediates have been identified, they have been the same as in pigeon liver, and precursor labeling experiments have led to similar results in all cases. On these grounds, therefore, this pathway apparently occurs in mammals, bacteria, fungi, plants, insects, and moluscs. The only known exception is the question, mentioned above, regarding the role of phosphoribosyl formylglycineamide synthetase in *Salmonella typhimurium*.

B. REACTION TYPES

Table 7–II lists the ten reactions of this pathway in terms of the general type of bond formed or broken, reversibility, and requirement for ATP.

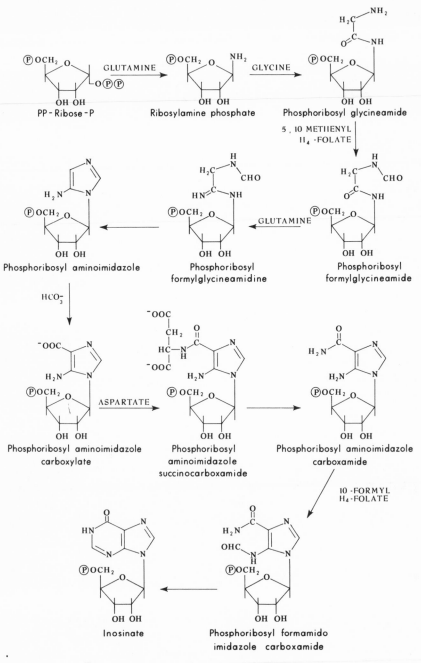

FIG. 7-2. Reactions of purine biosynthesis *de novo*.

TABLE 7-II

TYPES OF REACTIONS IN THE PATHWAY OF PURINE BIOSYNTHESIS *de Novo*

Reaction	Bond formed or broken	Reaction type	Reversible	ATP required
Phosphoribosyl pyrophosphate amidotransferase	Carbon–nitrogen	Glutamine amide transfer, phosphoribosyltransferase	No	No
Phosphoribosyl glycineamide synthetase	Carbon–nitrogen	Amide synthesis	Yes	Yes
Phosphoribosyl glycineamide formyltransferase	Carbon–nitrogen	One-carbon transfer	No	No
Phosphoribosyl formylglycineamidine synthetase	Carbon–nitrogen	Glutamine amide transfer	No	Yes
Phosphoribosyl aminoimidazole synthetase	Carbon–nitrogen	Cyclization	No	Yes
Phosphoribosyl aminoimidazole carboxylase	Carbon–carbon	CO_2 fixation	Yes	No
Phosphoribosyl aminoimidazole succinocarboxamide synthetase	Carbon–nitrogen	Aspartate amino transfer	Yes	Yes
Adenylosuccinate lyase	Carbon–nitrogen	Displacement	Yes	No
Phosphoribosyl aminoimidazole carboxamide formyltransferase	Carbon–nitrogen	One-carbon transfer	Yes	No
IMP cyclohydrolase	Carbon–nitrogen	Cyclization	Yes	No

Eight reactions involve carbon–nitrogen bond synthesis, in one a carbon–nitrogen bond is broken, and in only one case is a carbon–carbon bond formed. There are no oxidation–reduction reactions, and only one vitamin, folic acid, is involved. Four steps require ATP, but the first reaction uses PP-ribose-P as an equivalent energy donor. Four of the first five reactions are irreversible, whereas the last five are all reversible to some degree.

C. REVERSIBILITY

Under physiological conditions the pathway of purine biosynthesis *de novo* is believed to be irreversible. Reversibility of some reactions can, however, be demonstrated under some conditions of incubation *in vitro*. If Ehrlich ascites tumor cells are incubated with formate-^{14}C in the absence of glucose, for example, the 2-position of the purine ring may contain 8 to 10 times as much ^{14}C as does the 8-position (*30*), although these two positions should be equally labeled if net synthesis only had occurred. Apparently inosinate can be reversibly converted to phosphoribosyl aminoimidazole carboxamide, which in reforming inosinate incorporates radioactive formate.

IV. Control of Purine Biosynthesis *de Novo*

The term "control" is used here to include the wide variety of factors which may influence the rate of multienzyme reaction sequences.

A. ENZYME PATTERN

Potter has used this term to include the amount, activity, and localization of an enzyme. Little is known about the amount of any of the enzymes of the purine biosynthetic pathway, although bacterial mutants deficient in almost every one of these enzyme activities have been isolated. Mature mammalian erythrocytes also cannot synthesize purines *de novo*; the last two enzymes of this pathway are known to be present in these cells, but PP-ribose-P amidotransferase, the first enzyme, appears to be missing (*31*).

The term "enzyme activity" includes not only such kinetic parameters as Michaelis constants and maximum velocities, but also a consideration of the fraction of the potential activity which is available under the actual intracellular conditions of pH, temperature, ionic strength, etc. These factors cannot be evaluated for the reactions in question at the present time.

So far as is known, all of the enzymes of purine biosynthesis *de novo* are soluble and in the cytoplasmic fraction of cells. It has recently been suggested that in pigeon liver extracts the enzymes of the first four reactions in this pathway exist as a macromolecular complex, although this can be split into its component parts by appropriate treatments (*10*). The last two enzymes of the pathway also form a single larger unit in *Salmonella* (*29*).

B. AVAILABILIY OF SUBSTRATES

Regardless of the potential activity of the enzymes of purine biosynthesis *de novo* in any cell, the rates of the reactions they catalyze may be limited by the availability of substrates. The following questions may be asked: (a) Are the substrates required for purine biosynthesis made in, or are they available to, the cells in which this pathway operates? (b) Can such compounds readily gain entry to the cells? (c) Is the availability of these substrates for purine biosynthesis affected by alternative metabolic pathways for their utilization?

PP-Ribose-P is synthesized from glucose, and its rate of synthesis is affected both by the rate of glycolysis and by factors which affect the operation of the pentose phosphate cycle (see Chapter 6). Purine biosynthesis *de novo* must complete for this substrate with various phosphoribosyltransferases (see Chapter 5), and under some conditions these limit the availability of PP-ribose-P for this process.

PP-Ribose-P concentrations are much lower in Ehrlich ascites tumor cells *in vivo* than *in vitro*, and both PP-ribose-P levels and the rate of purine biosynthesis *de novo* are increased following the injection of glucose (*32*). Both PP-ribose-P concentrations and rates of *de novo* synthesis are increased markedly in human skin fibroblasts which are deficient in hypoxanthine-guanine phosphoribosyltransferase (*33*), and this is probably related to the extraordinary rate of purine synthesis *in vivo* in patients with this enzyme deficiency (*34*). PP-Ribose-P levels are also elevated in some patients with gout and accelerated purine biosynthesis *de novo* (*35*, *36*).

Purine biosynthesis *de novo* requires 4 moles of ATP per mole of inosinate formed, in addition to whatever is required for substrate and cofactor synthesis. It is usually assumed that ATP is present in sufficient concentration, but this has not been tested directly.

Glutamine is present in abundance in some tissues, such as brain, but is scarce in many tumor cells and is a rate-limiting factor in some cells, both *in vivo* and *in vitro*; it may even be limiting in liver under certain conditions (*17*, *37*). Similarly, glycine and aspartate can be rate-limiting for purine synthesis in some cells (*17*, *38*, *39*). As discussed above, aspartate can sometimes also become limiting in biotin deficiency. In addition to dietary supply, glycine and formate are together synthesized from glucose via serine by serine hydroxymethyltransferase (EC 2.1.2.1):

$$\text{Serine} + \text{H}_4\text{-folate} \rightleftharpoons \text{glycine} + \text{5,10-methylene H}_4\text{-folate}$$

CO_2 is presumed to be present from respiration and, to a lesser extent, from other decarboxylation reactions. Whether it becomes limiting under anaerobic conditions is not known.

The availability of Mg^{2+}, K^+, and the folate coenzymes may also be considered. The two cations are probably not rate-limiting, although K^+ concentrations do fluctuate in cells. Little is known about the concentrations of total folic acid or of individual folate coenzymes, relative to tissue requirements for them. That this cofactor may be in excess was suggested by a recent study (*39*) in which a 95% decrease in total H_4-folate was induced without causing any change in growth rate of the cells concerned. Others have found, however, that dietary deficiency of folate or vitamin B_{12} leads to increased urinary excretion of aminoimidazole carboxamide (*40, 41*).

C. CONTROL AT THE FIRST REACTION

The first reaction of purine biosynthesis *de novo*, PP-ribose-P amidotransferase, is generally believed to be an important point of regulation of the pathway as a whole. If purines are made available to cells, or injected into animals, purine biosynthesis *de novo* is inhibited, and the site of inhibition is known to be very early in the pathway. The purines must be converted to nucleotides to be inhibitory, and three modes of inhibition seem possible: (a) inhibition of PP-ribose-P amidotransferase by purine nucleotides (see below); (b) inhibition of PP-ribose-P synthesis by purine nucleotides (see Chapter 6); and (c) diversion of PP-ribose-P by using it in the conversion of purines to nucleotides (see Chapter 8). Although most attention is usually given to inhibition of PP-ribose-P amidotransferase by purine nucleotides, it must be stressed that the physiological significance of this phenomenon is not certain. Similarly, the site and mode of regulation of purine biosynthesis *de novo* in the absence of added purines is unclear.

1. *Inhibition of PP-Ribose-P Amidotransferase*

The kinetic parameters of PP-ribose-P amidotransferases are unexceptional (Section II B,1), and although limited PP-ribose-P and glutamine availability do play a role in the control of this enzyme, the imposition of control by end products of this pathway on the amount and activity of this enzyme is also believed to be important.

A variety of studies have demonstrated that the rate of PP-ribose-P amidotransferase activity is controlled by the concentration of purine nucleotides which are the end products of the *de novo* pathway. Wyngaarden first showed that PP-ribose-P amidotransferase is inhibited by adenine and guanine ribonucleotides, and inhibition of this enzyme by purine nucleotides has now been demonstrated in microbial, avian, and mammalian preparations.

PP-Ribose-P amidotransferases from various sources are inhibited by inosinate and by most adenine and guanine ribonucleotides, but the nucleotide which is most effective seems to differ from preparation to preparation. Although kinetic studies with pigeon liver enzyme have found that such inhibition is formally competitive with respect to PP-ribose-P and noncompetitive with respect to glutamine, it is apparent that inhibition is not of the classical sort. Inhibition is not competitive with respect to PP-ribose-P with a preparation from tumor cells (42), however. The observations of Nierlich and Magasanik (43) that combinations of various adenine and guanine ribonucleotides gave more inhibition than either did alone suggest that there are two separate inhibitor sites on the enzyme in addition to those for the substrates and that the binding of inhibitors results in conformational changes in the enzyme protein. There is some evidence that activation of PP-ribose-P amidotransferase by purine nucleotides occurs under some conditions and physiological control might be exerted by counteracting inhibitory and activating effects of purine nucleotides. The recent elucidation of the subunit structure of this enzyme by Rowe and Wyngaarden (10) may be the basis of the action of ribonucleotide feedback inhibitors. In the presence of PP-ribose-P the 200,000 molecular weight form is found and it is at least possible that inhibitors induce dissociation of this protein form into less active subunits. This remains to be demonstrated, however.

The relevance of these studies of the partially purified enzyme to the regulation of purine biosynthesis *in vivo* is uncertain.

Recent papers by Rowe and Wyngaarden (10), Shiio and Ishii (44), and Hill and Bennett (42) may be consulted for references, as well as the reviews by Stadtman (45) and Blakley and Vitols (46).

2. Repression

In addition to the end product control of enzyme activity discussed above, end product control of enzyme amount, or repression, is also possible. The latter has been studied much less than the former in this field, and the physiological significance of repression for the control of purine biosynthesis *de novo* is not clear. Nierlich and Magasanik (*44*) have reported that in *Aerobacter aerogenes*, PP-ribose-P amidotransferase and phosphoribosyl formylglycineamidine synthetase are coordinately repressed by addition of either adenine or guanine to the culture medium, and they were maximally repressed only when high levels of both bases were present. Phosphoribosyl glycineamide synthetase can also be repressed, but not to the same degree as these amidotransferases. The synthetase is also repressed slightly in animal cells grown in tissue culture in the presence of purines (*47*). The repression of PP-ribose-P amidotransferase, adenylosuccinate lyase, phosphoribosyl aminoimidazole carboxamide formyltransferase, and inosinate cyclohydrolase by the addition of either adenosine or guanosine to *Bacillus subtilis* cultures has also been reported (*48, 49*). The probable derepression of PP-ribose-P amidotransferase in bacteria (*50*) and mouse spleen infected with murine leukemia virus (*51*) has been observed.

The genetic loci in bacteria for the enzymes of various biosynthetic pathways are sometimes found to be closely linked, and repression may be expressed coordinately through a relatively simple operator system. The loci for the enzymes of purine biosynthesis *de novo* have recently been mapped in *S. typhimurium* and *E. coli*, and it is clear from the data shown in Table 7–III that they are scattered all over the bacterial chromosome, although a few clusters of two or three loci do occur (*29, 52, 53*).

TABLE 7–III

GENETIC LOCI FOR ENZYMES OF PURINE BIOSYNTHESIS *de Novo*

Enzyme	Map position	
	E. coli	*S. typhimurium*
Phosphoribosyl pyrophosphate amidotransferase	44	73
Phosphoribosyl glycineamide synthetase	78	129
Phosphoribosyl glycineamide formyltransferase	49	—
Phosphoribosyl formylglycineamidine synthetase	—	80
Phosphoribosyl aminoimidazole synthetase	—	79
Phosphoribosyl aminoimidazole carboxylase	15	19
Phosphoribosyl aminoimidazole succinocarboxamide synthetase	48	79
Adenylosuccinate lyase	23	43
Phosphoribosyl aminoimidazole carboxamide formyltransferase	78	129
IMP cyclohydrolase	—	129

D. FEEDBACK INHIBITION AND REPRESSION OF SUBSTRATE AND COFACTOR
SYNTHESIS

The end products of purine biosynthesis *de novo* can in some systems control the rate of synthesis of some of the substrates and cofactors of this pathway. In most cases these complex control mechanisms have been found in bacteria, but not in animal cells.

The glutamine synthetase of *E. coli* is partially inhibited by adenylate, being maximally inhibited only in the presence of products of eight different glutamine-utilizing pathways. The interpretation of these results is complicated by the more recent discovery of two forms of glutamine synthetase in this organism and of the two enzymes required for their interconversion (54). The sheep brain enzyme is not subject to these kinds of control mechanism (55).

Several of the enzymes which catalyze the interconversion of the H_4-folate coenzymes are inhibited or repressed by purine nucleotides. Thus, 5,10-methylene H_4-folate dehydrogenase is inhibited by ATP, GTP, and ITP (56), and 5,10-methylene H_4-folate reductase is inhibited by S-adenosylmethionine (57). 10-Formyl H_4-folate synthetase is repressed in cells grown in the presence of purines (58).

The functional significance of these phenomena remains to be evaluated.

E. CONTROL BY ANTIMETABOLITE ACTION

The effects of inhibitors may be considered to be one means by which the rate of purine biosynthesis may be controlled. Three different types of inhibitor actions may be distinguished.

Azaserine, diazo-oxo-norleucine, and some related compounds may be considered to be classical competitive inhibitors which act by reason of their structural resemblance to glutamine. Their action has already been discussed (Chapter 5).

A number of purine nucleotide analogues can mimic the effects of natural purine ribonucleotides on PP-ribose-P amidotransferase and thereby effect feedback inhibition of the entire pathway (59, 60). Such nucleotide analogues may be synthesized by cells from purine bases or nucleosides presented to them. A number of compounds with potent inhibitory properties, such as 6-mercaptopurine, 6-thioguanine, and 6-methylmercaptopurine ribonucleoside, inhibit purine biosynthesis *de novo* in this way, although it is not known whether this action is responsible for the growth inhibition.

Antimetabolites may also inhibit the synthesis of cofactors and substrates required for purine biosynthesis. As mentioned above (Chapter 6), PP-ribose-P synthesis can be inhibited by 3'-deoxyadenosine triphosphate, formycin triphosphate, and other ATP analogues (*59, 60*). Methionine sulfoximine can inhibit glutamine synthesis (*55*), and biotin analogues can inhibit aspartate formation. Folic acid analogues such as aminopterin and amethopterin, which are effective inhibitors of purine biosynthesis, act by inhibiting the reduction of folate to H_2-folate and H_4-folate. Such action leads not only to a shortage of one-carbon units, but may also cause a deficiency in glycine.

TABLE 7–IV

ENZYMES OF PURINE RIBONUCLEOTIDE BIOSYNTHESIS *de Novo*[a]

	Enzyme Commission number	Systematic name	Trivial name
1	2.4.2.14	Ribosylamine-5-phosphate: pyrophosphate phospho-ribosyltransferase (glutamate-amidating)	Phosphoribosyl pyrophosphate amidotransferase
2	6.3.1.3	Ribosylamine-5-phosphate: glycine ligase (ADP)	Phosphoribosyl glycineamide synthetase
3	2.1.2.2	5′-Phosphoribosyl-*N*-formyl-glycineamide: tetrahydro-folate 5,10-formyltransferase	Phosphoribosyl glycineamide formyltransferase
4	6.3.5.3	5′-Phosphoribosyl-formyl-glycineamide: L-glutamine amido-ligase (ADP)	Phosphoribosyl formyl-glycineamidine synthetase
5	6.3.3.1	5′-Phosphoribosyl-formyl-glycineamidine cyclo-ligase (ADP)	Phosphoribosyl aminoimidazole synthetase
6	4.1.1.21	5′-Phosphoribosyl-5-amino-4-imidazolecarboxylate carboxylyase	Phosphoribosyl aminoimidazole carboxylase
7	6.3.2.6	5′-Phosphoribosyl-4-carboxy-5-aminoimidazole: L-aspartate ligase (ADP)	Phosphoribosyl aminoimidazole succinocarboxamide synthetase
8	4.3.2.2	Adenylosuccinate AMP-lyase	Adenylosuccinate lyase
9	2.1.2.3	5′-Phosphoribosyl-5-form-amide-4-imidazolecarbox-amide: tetrahydrofolate 10-formyltransferase	Phosphoribosyl aminoimidazole carboxamide formyl-transferase
10	3.5.4.10	IMP 1,2-hydrolase (decyclizing)	IMP cyclohydrolase

[a] Numbers in the first column refer to reactions in the summary diagram (p. 121).

V. Physiological Significance

Cells which do not obtain preformed purine bases or nucleosides from their environment must make purines *de novo* in order to live. The pathway of purine synthesis *de novo* is therefore essential for a wide variety of microorganisms and for plant and animal cells in purine-free environments.

Conversely, some cells have an absolute requirement for exogenously supplied purines because of a deficiency of one or more of the enzymes in the pathway of synthesis *de novo*.

Many cells, including most animal cells *in vivo*, fall somewhere between these two groups and meet part of their purine requirements by synthesis *de novo* and part through utilization of exogenous purines. In such cases it is difficult to determine exactly how much of their purine content is derived from each source under physiological conditions.

VI. Summary

The pathway of purine ribonucleotide biosynthesis *de novo* is summarized below. The enzymes involved are listed in Table 7–IV.

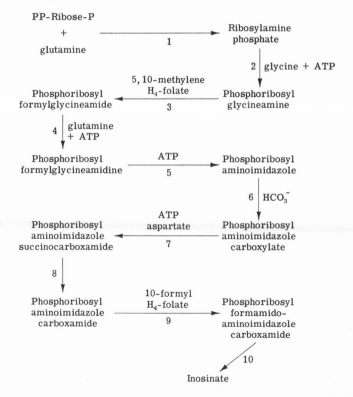

References

1. McCrudden, F. H., "Uric Acid." Fort Hill Press, Boston, Massachusetts, 1905.
2. Scheele, K. H., "Chemical Essays" (Engl. transl.). Scott, Greenwood, London, 1901.
3. Buchanan, J. M., and Hartman, S. C., *Advan. Enzymol.* **21**, 200 (1959).
4. Buchanan, J. M., *in* "The Nucleic Acids" (E. Chargaff and J. N. Davidson, eds.), Vol. 3, p. 304. Academic Press, New York, 1960.
5. Hartman, S. C., and Buchanan, J. M., *Ergeb. Physiol., Biol. Chem. Exp. Pharmakol.* **50**, 75 (1959).
6. Moat, A. G., and Friedman, H., *Bacteriol. Rev.* **24**, 309 (1960).
7. Magasanik, B., *in* "The Bacteria" (I. C. Gunsalus and R. Y. Stanier, eds.), Vol 3, p. 295. Academic Press, New York, 1962.
8. Greenberg, G. R., *J. Biol. Chem.* **190**, 611 (1951).
9. Hartman, S. C., *J. Biol. Chem.* **238**, 3024 (1963).
10. Rowe, P. B., and Wyngaarden, J. B., *J. Biol. Chem.* **243**, 6373 (1968).
11. Hartman, S. C., *J. Biol. Chem.* **238**, 3036 (1963).
12. Wyngaarden, J. B., and Ashton, D. M., *J. Biol. Chem.* **234**, 1492 (1959).
13. Nierlich, D. P., and Magasanik, B., *J. Biol. Chem.* **240**, 366 (1965).
14. Herscovics, A., and Johnstone, R. M., *Biochim. Biophys. Acta* **93**, 251 (1964).
15. Trachewsky, D., and Johnstone, R. M., *Can. J. Biochem.* **47**, 839 (1969).
16. Le Gal, M. L., Le Gal, Y., Roche, J., and Hedegaard, J., *Biochem. Biophys. Res. Commun.* **27**, 618 (1962).
17. Fontenelle, L. J., and Henderson, J. F., *Biochim. Biophys. Acta* **177**, 88 (1969).
18. Reem, G. H., *J. Biol. Chem.* **243**, 5695 (1968).
19. Westby, C. A., and Gots, J. S., *J. Biol. Chem.* **244**, 2095 (1969).
20. French, T. C., Dawid, I. B., Day, R. A., and Buchanan, J. M., *J. Biol. Chem.* **238**, 2171 (1963).
21. Mizobuchi, K., and Buchanan, J. M., *J. Biol. Chem.* **243**, 4842 (1968).
22. Mizobuchi, K., and Buchanan, J. M., *J. Biol. Chem.* **243**, 4853 (1968).
23. Mizobuchi, K., Kenyon, G. L., and Buchanan, J. M., *J. Biol. Chem.* **243**, 4863 (1968).
24. Schroeder, D. D., Allison, A. J., and Buchanan, J. M., *J. Biol. Chem.* **244**, 5856 (1969).
25. Dawid, I. B., French, T. C., and Buchanan, J. M., *J. Biol. Chem.* **238**, 2178 (1963).
26. Ahmad, F., Missimer, P., and Moat, A. G., *Can. J. Biochem.* **43**, 1723 (1965).
27. Meister, A., "Biochemistry of the Amino Acids," 2nd ed., Vols. 1 and 2. Academic Press, New York, 1965.
28. Fisher, C. R., *Biochim. Biophys. Acta* **178**, 380 (1969).
29. Gots, J. S., Dalal, F. R., and Shumas, S. R., *J. Bacteriol.* **99**, 441 (1969).
30. Smellie, R. M. S., Thomson, R. Y., Goutier, R., and Davidson, J. N., *Biochim. Biophys. Acta* **22**, 585 (1956).
31. Fontenelle, L. J., and Henderson, J. F., *Biochim. Biophys. Acta,* **177**, 175 (1969).
32. Henderson, J. F., and Khoo, M. K. Y., *J. Biol. Chem.* **240**, 2349 (1965).
33. Rosenbloom, F. M., Henderson, J. F., Caldwell, I. C., Kelley, W. N., and Seegmiller, J. E., *J. Biol. Chem.* **243**, 1166 (1968).
34. Lesch, M., and Nyhan, W. L., *Amer. J. Med.* **36**, 561 (1964).
35. Hershko, A., Hershko, C., and Mager, J., *Isr. J. Med. Sci.* **4**, 939 (1968).
36. Henderson, J. F., Rosenbloom, F. M., Kelley, W. M., and Seegmiller, J. E., *J. Clin. Invest.* **47**, 1511 (1968).
37. Feigelson, M., and Feigelson, P., *J. Biol. Chem.* **241**, 5819 (1966).

38. Bloomfield, R. A., Letter, A. A., and Wilson, R. P., *Arch. Biochem. Biophys.* **129**, 196 (1969).
39. Sotobaishi, H., Rosen, F., and Nichol, C. A., *Biochemistry* **5**, 3879 (1966).
40. McGeer, P. L., Sen, N. P., and Grant, D. A., *Can. J. Biochem.* **43**, 1367 (1965).
41. Oace, S. M., Tarczy-Hornoch, K., and Stokstad, E. L. R., *J. Nutr.* **95**, 445(1968).
42. Hill, D. L., and Bennett, L. L., Jr., *Biochemistry* **8**, 122 (1969).
43. Nierlich, D. P., and Magasanik, B., *J. Biol. Chem.* **240**, 358 (1965).
44. Shiio, I., and Ishii, K., *J. Biochem.* (*Tokyo*) **66**, 175 (1969).
45. Stadtman, E. R., *Advan. Enzymol.* **28**, 41 (1966).
46. Blakley, R. L., and Vitols, E., *Annu. Rev. Biochem.* **37**, 201 (1968).
47. Nierlich, D. P., and McFall, E., *Biochim. Biophys. Acta* **76**, 469 (1963).
48. Nishikawa, H., Momose, H., and Shiio, I., *J. Biochem.* (*Tokyo*) **62**, 92 (1967).
49. Momose, H., Nishikawa, H., and Shiio, I., *J. Biochem.* (*Tokyo*) **59**, 325 (1966).
50. Love, S. H., and Remy, C. N., *J. Bacteriol.* **91**, 1037 (1966).
51. Reem, G. H., and Friend, C., *Science* **157**, 1203 (1967).
52. Sanderson, K. E., *Bacteriol. Rev.* **31**, 354 (1967).
53. Taylor, A. L., and Trotten, C. D., *Bacteriol. Rev.* **31**, 332 (1967).
54. Stadtman, E. R., Shapito, B. M., Kingdom, H. S., Woolfolk, C. A., and Hubbard, J. S., *Advan. Enzyme Regul.* **6**, 257 (1968).
55. Meister, A., *Harvey Lect.* **63**, 139 (1969).
56. Dalal, F. R., and Gots, J. S., *J. Biol. Chem.* **242**, 3636 (1967).
57. Kutzbach, C., and Stokstad, E. L. R., *Biochim. Biophys. Acta* **139**, 217 (1967).
58. Albrecht, A. M., and Hutchison, D. J., *J. Bacteriol.* **87**, 792 (1964).
59. Henderson, J. F., *Progr. Exp. Tumor Res.* **6**, 84 (1965).
60. Balis, M. E., "Antagonists of Nucleic Acids." North-Holland Publ., Amsterdam, 1968.

CHAPTER 8

PURINE RIBONUCLEOTIDE SYNTHESIS FROM PURINE BASES AND RIBONUCLEOSIDES

I. Introduction and Early History

Although the significance of purine biosynthesis *de novo* was appreciated many years ago, it was not known until relatively recently that cells could also utilize purine bases and nucleosides. This possibility had indeed been studied by early investigators, but because of the limitations of their methods, the results obtained remained inconclusive or contradictory. This work has been reviewed by Rose (1) and by Christman (2).

The fact that purine bases could be used directly for nucleotide and nucleic acid synthesis was first established by the use of labeled compounds. Plentl and Schoenheimer (3) were not able to demonstrate incorporation of ^{15}N-guanine into nucleic acids of rat viscera in 1944, but Brown later was able to show that it was utilized by mice. (It is now known that guanine is degraded more rapidly in rats than in mice.) Brown and his colleagues (4, 5) also demonstrated the incorporation of ^{15}N-adenine into nucleic acids, and between 1948 and 1954 numerous studies were made of the incorporation of ^{15}N- or ^{14}C-labeled purines into nucleotides and nucleic acids (see reference 2).

The physiological significance of nucleotide and nucleic acid synthesis from purine bases was demonstrated by three types of evidence: (a) microorganisms were isolated that required one or more purine bases for growth (6, 7), (b) inhibition of purine biosynthesis de novo with certain anticancer drugs resulted in only partial inhibition of tumor growth, (c) genetic abnormalities in one of the enzymes of purine utilization led to severe neuropathy in children (8). The relative importance of purine base utilization and purine biosynthesis de novo varies greatly from one type of cell to another, however.

Purine bases can be converted to ribonucleotides via phosphoribosyl-transferases; PP-ribose-P provides the ribosyl phosphate moiety. Purine nucleosides can be phosphorylated by ATP-requiring nucleoside kinases to form the same ribonucleotides. Finally, the possibility also exists that purine bases are first converted to ribonucleosides via nucleoside phosphorylase, and then to ribonucleotides by the above-mentioned kinases. These routes of ribonucleotide synthesis are summarized as follows:

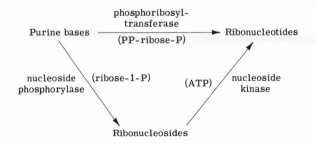

II. Purine Phosphoribosyltransferases

A. INTRODUCTION

It was not until 1953 that Goldwasser (9) and Williams and Buchanan (10) showed that purine bases could be converted to ribonucleotides by a one-step process, without the intermediate formation of ribonucleosides. The source of the ribose phosphate moiety was discovered in 1955 to be PP-ribose-P in the course of studies of adenylate synthesis by Kornberg et al. (11), and of inosinate synthesis by Korn et al. (12); extracts of yeast, beef liver, and pigeon liver were employed. The enzymes involved were at first called nucleotide pyrophosphorylases, but are now known as purine phosphoribosyltransferases. The general reaction is

$$\text{Purine base} + \text{PP-ribose-P} \xrightarrow{\text{Mg}^{2+}} \text{purine ribonucleotide} + \text{PP}_i$$

Further studies have shown that there are two important purine phosphoribosyltransferases in animal tissues and three in some microorganisms. This topic has recently been exhaustively reviewed by Murray et al. (13), Raivio and Seegmiller (14), and by Murray (15).

B. ADENINE PHOSPHORIBOSYLTRANSFERASE

The existence of a separate phosphoribosyltransferase for adenine was first deduced because gentle heating of crude enzyme preparations destroyed this activity, without decreasing their ability to make inosinate from hypoxanthine and PP-ribose-P. Adenine phosphoribosyltransferases have since been separated from other purine phosphoribosyltransferases by ion-exchange chromatography, electrophoresis, and gel filtration, and the separate identities of these enzymes have also been demonstrated by specific mutations, both in man and in bacteria.

Adenine phosphoribosyltransferase has strict requirements for binding of the purine base, and even small changes in purine structure markedly reduce the ability of an analogue to serve as substrate or competitive inhibitor (16, 17). It is believed that adenine binds to the enzyme through the 6-amino group and the 3- and 7-nitrogens. The binding of PP-ribose-P requires that the anionic phosphate groups be separated by a specific distance; the pyrophosphate moiety must also be complexed with Mg^{2+} (17, 18). The ribose can be replaced by other sugars or even by a hydrocarbon chain. The binding of PP-ribose-P appears to cause conformational changes in the enzyme which may strain the glycosidic bond and thereby facilitate catalysis, and the rate-limiting step in the reaction is the removal of the hydrogen from N-9 (19).

Kinetic studies have shown that the reaction is ordered, that ternary enzyme–substrate complexes are involved, and that PP-ribose-P binds first (20–22); adenine therefore does not bind to free enzyme (23, 24).

Michaelis constants of about 0.8 and 6 μM for adenine and PP-ribose-P, respectively, for a tumor cell enzyme (24, 25), and 2 to 140 μM and 62 to 64 μM for the human erythrocyte enzyme (22, 26), not only ensure that this enzyme has a great competitive advantage for PP-ribose-P, a substrate common to a number of enzymes, but may also help to explain why free adenine virtually does not exist in tissues.

The tissue distribution of adenine phosphoribosyltransferase varies considerably in rat, mouse, man, and monkey, and representative data are shown in Table 8–I. Murray (30) has found that the kinetic parameters of this enzyme in mouse liver vary during the course of embryonic development and has suggested that two forms of the enzyme may appear in succession. Both Murray (30) and Epstein (31) have observed marked changes in adenine phosphoribosyltransferase activity during embryonic development.

Genetically altered adenine phosphoribosyltransferases have been found in man (32, 33) and in *Salmonella* (34, 35). One type of mutation in man leads to the presence of about 20% of normal activity, while others lead to increased activity, or to variable stability to heating. Both types of mutation are inherited in an autosomal manner.

TABLE 8–I

Tissue Distribution of Adenine
Phosphoribosyltransferase in
Different Species

Tissue	AMP formed (mμmole/min/mg protein)			
	Mouse[a]	Rat[a]	Man[b]	Monkey[c]
Liver	0.15	0.38	2.6	21.7
Brain	0.08	0.30	1.1	3.2
Spleen	0.18	1.31	0.83	11.2
Kidney	0.14	0.46	1.18	9.9

[a] From Murray (27).
[b] From Rosenbloom *et al.* (28).
[c] From Krenitsky (29).

C. HYPOXANTHINE-GUANINE PHOSPHORIBOSYLTRANSFERASE

In mammalian cells, yeasts, and some bacteria (e.g., *Lactobacillus casei*), a single enzyme catalyzes the reactions of hypoxanthine and guanine with PP-ribose-P. Both physical and kinetic evidence support the view that the two purine bases share the same binding site, as do inosinate and guanylate. The hypoxanthine-guanine phosphoribosyltransferase from mammalian cells and yeasts apparently also catalyzes nucleotide formation from xanthine, whereas separate enzymes for xanthine exist in some bacteria (see below). Its substrates also include such purine analogues as 6-mercaptopurine, 6-thioguanine, and 8-azaguanine. This enzyme is remarkably stable to heating, and molecular weights of preparations from various sources range from 44,000 (36) to 80,000 (37).

Like adenine phosphoribosyltransferase, that for hypoxanthine and guanine catalyzes an ordered reaction; PP-ribose-P binds to form a ternary complex (38). Krenitsky and Papaioannou (39), however, suggest the possibility that alternative reaction sequences may also be involved. The PP-ribose-P reacts in the form of a Mg^{2+} complex.

The Michaelis constants of the enzyme from various sources are about 4 and 10 μM for guanine and hypoxanthine, respectively, and 6 to 200 μM for PP-ribose-P (25, 36, 37, 38, 40). In all cases the maximum velocity with hypoxanthine is about 60% of that using guanine.

The relative and total activities of hypoxanthine-guanine phosphoribosyltransferases in different tissues varies from species to species (Table 8–II), as does that of adenine phosphoribosyltransferase. In man and

TABLE 8–II

TISSUE DISTRIBUTION OF HYPOXANTHINE
PHOSPHORIBOSYLTRANSFERASE IN
DIFFERENT SPECIES

Tissue	AMP formed (mμmole/min/mg protein)			
	Mouse[a]	Rat[a]	Man[b]	Monkey[c]
Liver	0.25	0.44	0.69	8.6
Brain	0.95	0.52	8.3	55.5
Spleen	0.53	0.47	0.60	15.1
Kidney	0.20	0.36	0.47	7.6

[a] From Murray (27).
[b] From Rosenbloom et al. (28).
[c] From Krenitsky (29).

monkey the activity of this enzyme is highest in brain; the human basal ganglia in particular have more activity than other parts of this organ.

Several types of human mutants with hypoxanthine-guanine phosphoribosyltransferase defects have recently been identified (8, 41, 42). Cells from patients with the so-called Lesch-Nyhan syndrome have very low (43) or undetectable activities of this enzyme, but immunological studies have demonstrated that the enzyme protein is made even when it is catalytically inactive (44). In some patients with gout, 0.05 to 10% of normal activity remains; isozymes (45) and other mutants (46) have also been detected. The structural gene for hypoxanthine-guanine phosphoribosyltransferase is on the X chromosome, and mutations at this locus are, therefore, only fully expressed in males.

D. OTHER PURINE PHOSPHORIBOSYLTRANSFERASES

A separate xanthine phosphoribosyltransferase has been shown to be present in *Salmonella typhimurium* (34, 47) and in *L. casei* (36). Krenitsky *et al.* (36) have also recently demonstrated that in *Escherichia coli* there is a separate hypoxanthine phosphoribosyltransferase, and that guanine and xanthine phosphoribosyltransferase activities are present in a single enzyme.

Finally, the uracil phosphoribosyltransferase of beef erythrocytes recognizes the 2,4-dioxypyrimidine moiety of xanthine and uric acid, and attaches the phosphoribosyl group to the 3-position of these purines, as it is analogous with the N-1 of uracil. Uric acid 3-ribonucleoside is found as a constituent of beef erythrocytes, and presumably arises by dephosphorylation of the nucleotide derivative (48, 49).

E. PHYSIOLOGICAL ROLES

The purine phosphoribosyltransferases permit cells to use exogenous or dietary purines, and this function is undoubtedly important to some bacteria and to those animal cells (e.g., erythrocytes) which do not synthesize purines *de novo*. The importance of this role in other animal cells *in vivo* is far from clear, however, as hypoxanthine is present in serum only at very low concentrations (50), and adenine and guanine have not been detected in normal serum. Purine bases (especially hypoxanthine and guanine) can be produced intracellularly by the catabolism of messenger RNA and soluble purine nucleotides; their reutilization via the phosphoribosyltransferases would prevent the loss of these compounds from the cells.

Purine phosphoribosyltransferases may not be essential in all animal cells. Thus, experimental tumors *in vivo* and cells in culture can grow at normal rates without them. The best evidence for the importance of hypoxanthine-guanine phosphoribosyltransferase is that male children who completely lack this enzyme have a severe neurological disease characterized by self-destructive biting, mental retardation, spasticity, and choreoathetosis (*8, 51, 52*). Males who retain a few percent of normal enzyme activity usually develop gout (*41*).

F. PHARMACOLOGICAL ROLES

Whatever their physiological role, the purine phosphoribosyltransferases play important pharmacological roles through their "activation" of drugs for cancer chemotherapy and suppression of the immune response. Purine base analogues used for these purposes act after conversion to the nucleoside monophosphate, which may be the pharmacologically active form or an intermediate in the formation of the active form. For example, adenine phosphoribosyltransferase activates 2,6-diaminopurine, and hypoxanthine-guanine phosphoribosyltransferase activates 6-mercaptopurine, 6-thioguanine, and 8-azaguanine. Cells which lack the appropriate enzyme are insensitive or resistant to the corresponding drugs (*53, 54*).

III. Purine Ribonucleoside Kinases

Historically, the first route of purine nucleotide synthesis to be studied in detail was that involving the phosphorylation of adenosine by ATP. More recently, evidence has been presented for the existence in animal tissues of two other purine nucleoside kinases. The general reaction is

$$\text{Purine ribonucleoside} + \text{ATP} \xrightarrow{\text{kinase}} \text{purine ribonucleotide} + \text{ADP}$$

A. ADENOSINE KINASE

The adenosine kinases from animal cells have broad specificities with respect to their substrates. Thus, ATP, ITP, and GTP can all serve as phosphate donors, although to varying degrees. The nucleoside triphosphate is probably involved as the Mg^{2+} complex, and the Mg^{2+}-nucleoside triphosphate ratio is critical. The Michaelis constant for ATP is about $5-10 \times 10^{-4} M$, depending on the Mg^{2+} concentration. The molecular weight of the rabbit liver enzyme is about 233,000.

Although yeast adenosine kinase is relatively specific with respect to its purine nucleoside substrate (*55*), the enzymes from animal tissues phosphorylate a variety of nucleosides, in which either purine base or sugar moieties, or both, can be quite different from adenosine. The maximum velocity of phosphorylation of 6-methylmercaptopurine ribonucleoside is greater than that of adenosine; adenosine, however, has a Michaelis constant of 1.6×10^{-6} M, whereas that for 6-methylmercaptopurine ribonucleoside is 5×10^{-5} M (56–58).

B. INOSINE AND GUANOSINE KINASES

Studies of purified adenosine kinases have made it clear that inosine and guanosine are not among the substrates of this enzyme, in spite of its broad specificity, and for many years it was not certain whether they were enzymatically phosphorylated or not. Recently, however, LePage and his co-workers (59, 60) have provided data supporting the existence of separate inosine and guanosine kinases in crude extracts of mouse tumors, but these activities have not yet been purified.

C. PHYSIOLOGICAL ROLES

The physiological role of the purine nucleoside kinases remains obscure. Tumor cells lacking adenosine kinase grow at normal rates both *in vivo* (61) and in tissue culture (62).

D. PHARMACOLOGICAL ROLES

Regardless of the physiological role of the purine nucleoside kinases, adenosine kinase is very important in cancer chemotherapy with purine nucleoside analogues. Attempts are being made to treat tumor cells with purine nucleosides in which the base and sugar moieties have been altered from the normal structures. Conversion to nucleotides is necessary for the expression of the pharmacological activity of many of these compounds, and this is accomplished through the broad substrate specificity of adenosine kinase.

IV. Anabolic Role of Purine Nucleoside Phosphorylase

The possibility has been considered that purine bases can be converted to ribonucleosides by purine nucleoside phosphorylase, and thence to ribonucleotides by the purine nucleoside kinases.

$$\text{Purine base} \underset{\text{P}_i,\ \text{phosphorylase}}{\overset{\text{ribose–1–P}}{\rightleftharpoons}} \text{ribonucleoside} \overset{\text{ATP}}{\underset{\text{kinase}}{\longrightarrow}} \text{ribonucleotide}$$

Although purine nucleoside phosphorylase is known to function in the catabolism of purine nucleosides (see Chapter 10), there is in fact no evidence that it functions as indicated in this scheme. Thus, in cells which lack purine phosphoribosyltransferases, purine bases are not converted to ribonucleotides despite the presence of phosphorylase and kinase activities.

V. Phosphate Transfer Pathway

In 1954 Chargaff (63) and associates discovered an enzyme (or group of enzymes) which was widely distributed in nature and which would transfer phosphate from a low-energy phosphate ester to a nucleoside.

$$\text{ROP} + \text{nucleoside} \rightarrow \text{nucleotide} + \text{ROH}$$

There was little specificity for the phosphate donor, and a number of nucleotides would serve as well as phenyl phosphate. A variety of purine and pyrimidine nucleosides act as phosphate acceptors, although the enzyme from carrot preferred pyrimidine nucleosides and would not phosphorylate adenosine (64). Other alcohols, including sugars, were not substrates for this enzyme.

Although this enzyme activity is widely distributed in nature (63), there is at present no clear evidence as to its role, or that it is an important route of nucleotide synthesis.

VI. Regulation of Purine Nucleotide Synthesis

Stadtman (65) and Blakley and Vitols (66) have reviewed studies of the inhibition and stimulation of the purine phosphoribosyltransferases by purine ribo- and deoxyribonucleotides. At relatively high concentrations, a variety of nucleotides inhibit these enzymes, while a few increase these activities at low concentrations. Studies by Henderson et al. (67) have shown that inhibitors bind to several kinetically significant forms of adenine phosphoribosyltransferase and that there are probably several different inhibitor binding sites which are not the same as those to which substrates and products bind. It must be emphasized, however, that the physiological significance of the studies conducted so far is unclear. Attempts to study the control of these reactions in intact cells have merely emphasized the complexity of this control. Under some conditions the availability of PP-ribose-P may limit the rate of these reactions.

Murray (58) has shown that a variety of purine nucleotides inhibit adenosine kinase, but the physiological significance of these observations remains to be established.

TABLE 8–III

ENZYMES OF PURINE RIBONUCLEOTIDE SYNTHESIS FROM BASES AND RIBONUCLEOSIDES

Enzyme Commission number	Systematic name	Trivial name
2.4.2.7	AMP:pyrophosphate phosphoribosyltransferase	Adenine phosphoribosyltransferase
2.4.2.8	IMP:pyrophosphate phosphoribosyltransferase	Hypoxanthine-guanine phosphoribosyltransferase
—	—	Xanthine phosphoribosyltransferase
—	—	2,4-Dioxypyrimidine phosphoribosyltransferase
2.7.1.20	ATP:adenosine 5'-phosphotransferase	Adenosine kinase
—	—	Inosine kinase
—	—	Guanosine kinase
2.4.2.1	Purine-nucleoside:orthophosphate ribosyltransferase	Purine nucleoside phosphorylase
—	—	Nucleoside phosphotransferase

VII. Summary

The pathways by which purine ribonucleotides are synthesized from purine bases and ribonucleosides are summarized below. The enzymes are listed in Table 8–III.

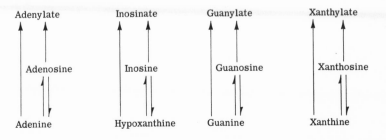

References

1. Rose, W. C., *Physiol. Rev.* **3**, 544 (1923).
2. Christman, A. C., *Physiol. Rev.* **32**, 303 (1952).
3. Plentl, A. A., and Schoenheimer, R., *J. Biol. Chem.* **153**, 203 (1944).
4. Brown, G. B., Bendich, H., Roll, P. M., and Suguera, F., *Proc. Soc. Exp. Biol. Med.* **72**, 501 (1949).
5. Brown, G. B., Roll, P. M., Plentl, A. A., and Cavalieri, L. F., *J. Biol. Chem.* **172**, 469 (1948).
6. Magasanik, B., *Annu. Rev. Microbiol.* **11**, 221 (1957).
7. Magasanik, B., *in* "The Bacteria" (I. C. Gunsalus and R. Y. Stanier, eds.), Vol. **3**, 295. Academic Press, New York, 1962.
8. Seminars on the Lesch-Nyhan Syndrome, *Fed. Proc., Fed. Amer. Soc. Exp. Biol.* **27**, 1019 (1968).
9. Goldwasser, E., *Biochim. Biophys. Acta* **13**, 341 (1954).
10. Williams, W. J., and Buchanan, J. M., *J. Biol. Chem.* **203**, 583 (1953).
11. Kornberg, A., Lieberman, I., and Simms, E. S., *J. Biol. Chem.* **215**, 417 (1955).
12. Korn, E. D., Remy, C. N., Wasilejko, H. C., and Buchanan, J. M., *J. Biol. Chem.* **217**, 875 (1955).
13. Murray, A. W., Elliott, D. C., and Atkinson, M. R., *Progr. Nucl. Acid Res. Mol. Biol.* **10**, 87 (1970).
14. Raivio, K. O., and Seegmiller, J. E., *Curr. Top. Enzyme Regul.* **2**, 201 (1970).
15. Murray, A. W., *Annu. Rev. Biochem.* **40**, 811 (1971).
16. Gadd, R. E. A., and Henderson, J. F., *Can. J. Biochem.* **48**, 295 (1970).
17. Krenitsky, T. A., Neil, S. A., Elion, G. B., and Hitchings, G. H., *J. Biol. Chem.* **244**, 4779 (1969).
18. Gadd, R. E. A., and Henderson, J. F., *J. Biol. Chem.* **245**, 2979 (1970).
19. Gadd, R. E. A., and Henderson, J. F., *Biochem. Biophys. Res. Commun.* **38**, 363 (1970).
20. Gadd, R. E. A., and Henderson, J. F., *Biochim. Biophys. Acta* **191**, 735 (1969).
21. Berlin, R. D., *Arch. Biochem. Biophys.* **134**, 120 (1969).
22. Srivastava, S. K., and Beutler, E., *Arch. Biochem. Biophys.* **142**, 426 (1971).
23. Stonehill, E. H., and Balis, M. E., *Anal. Biochem.* **10**, 386 (1965).
24. Hori, M., and Henderson, J. F., *J. Biol. Chem.* **241**, 3404 (1966).
25. Atkinson, M. R., and Murray, A. W., *Biochem. J.* **94**, 64 (1965).
26. Dean, D. M., Watts, T. W. E., and Westwick, W. J., *FEBS Lett.* **1**, 179 (1968).
27. Murray, A. W., *Biochem. J.* **100**, 664 (1966).
28. Rosenbloom, F. M., Kelley, W. N., Miller, J., Henderson, J. F., and Seegmiller, J. E., *J. Amer. Med. Ass.* **202**, 175 (1967).
29. Krenitsky, T. A., *Biochim. Biophys. Acta* **179**, 506 (1969).
30. Murray, A. W., *Biochem. J.* **104**, 675 (1967).
31. Epstein, C. J., *J. Biol. Chem.* **245**, 3289 (1970).
32. Kelley, W. N., Levy, R. I., Rosenbloom, F. M., Henderson, J. F., and Seegmiller, J. E., *J. Clin. Invest.* **47**, 2281 (1968).
33. Henderson, J. F., Kelley, W. H., Rosenbloom, F. M., and Seegmiller, J. E., *Amer. J. Hum. Genet.* **21**, 61 (1969).
34. Kalle, G. P., and Gots, J. S., *Science* **142**, 680 (1963).
35. Adye, J. C., and Gots, J. S., *Biochim. Biophys. Acta* **118**, 344 (1966).
36. Krenitsky, T. A., Neil, S. N., and Miller, R. L., *J. Biol. Chem.* **245**, 2605 (1970).

37. Krenitsky, T. A., Papaioannou, R., and Elion, G. B., *J. Biol. Chem.* **244,** 1263 (1969).
38. Henderson, J. F., Brox, L. W., Kelley, W. N., Rosenbloom, F. M., and Seegmiller, J. E., *J. Biol. Chem.* **243,** 2514 (1968).
39. Krenitsky, T. A., and Papaioannou, R., *J. Biol. Chem.* **244,** 1271 (1969).
40. Miller, J., and Beiber, S., *Biochemistry* **7,** 1420 (1968).
41. Kelley, W. N., Greene, M. L., Rosenbloom, F. M., Henderson, J. F., and Seegmiller, J. E., *Ann. Intern. Med.* **70,** 155 (1969).
42. Henderson, J. F., *Clin. Biochem.* **2,** 241 (1969).
43. McDonald, J. A., and Kelley, W. A., *Science* **171,** 689 (1971).
44. Rubin, C. S., Dancis, J., Yip, L. C., Nowinski, R. C., and Balis, M. E., *Proc. Nat. Acad. Sci. U.S.* **68,** 1461 (1971).
45. Bakay, B., and Nyhan, W. L., *Biochem. Genet.* **5,** 81 (1971).
46. Kelley, W. N., and Meade, J. C., *J. Biol. Chem.* **246,** 2953 (1971).
47. Kalle, G. P., and Gots, J. S., *Fed. Proc., Fed. Amer. Soc. Exp. Biol.* **20,** 358 (1961).
48. Hatfield, D., and Wyngaarden, J. B., *J. Biol. Chem.* **239,** 2580 (1964).
49. Hatfield, D., and Wyngaarden, J. B., *J. Biol. Chem.* **239,** 2587 (1964).
50. Goldfinger, S., Klinenberg, J. R., and Seegmiller, J. E., *J. Clin. Invest.* **44,** 623 (1967).
51. Lesch, M., and Nyhan, W. L., *Amer. J. Med.* **36,** 561 (1964).
52. Seegmiller, J. E., Rosenbloom, F. M., and Kelley, W. N., *Science* **155,** 1682 (1967).
53. Balis, M. E., "Antagonists and Nucleic Acids," p. 40. North-Holland Publ., Amsterdam, 1968.
54. Brockman, R. W., *Cancer Res.* **25,** 1596 (1965).
55. Greenberg, G. R., *J. Biol. Chem.* **219,** 423 (1956).
56. Lindberg, B., Klenow, H., and Hansen, K., *J. Biol. Chem.* **242,** 350 (1967).
57. Schnebli, H. P., Hill, D. L., and Bennett, L. L., Jr., *J. Biol. Chem.* **242,** 1997 (1967).
58. Murray, A. W., *Biochem. J.* **106,** 549 (1968).
59. Pierre, K. J., Kimball, A. P., and LePage, G. A., *Can. J. Biochem.* **45,** 1619 (1967).
60. Pierre, K. J., and LePage, G. A., *Proc. Soc. Exp. Biol. Med.* **127,** 432 (1968).
61. Caldwell, I. C., Henderson, J. F., and Paterson, A. R. P., *Can. J. Biochem.* **45,** 735 (1967).
62. Bennett, L. L., Jr., Schnebli, H. P., Vail, M. H., Allan, P. W., and Montgomery, J. A., *Mol. Pharmacol.* **2,** 432 (1966).
63. Brawerman, G., and Chargaff, E., *Biochim. Biophys. Acta* **15,** 558 (1954).
64. Tunis, M., and Chargaff, E., *Biochim. Biophys. Acta* **37,** 257 (1960).
65. Stadtman, E. R., *Advan. Enzymol.* **28,** 41 (1966).
66. Blakley, R. L., and Vitols, E., *Annu. Rev. Biochem.* **37,** 201 (1968).
67. Henderson, J. F., Gadd, R. E. A., Hori, M., and Miller, H., *Can. J. Biochem.* **48,** 573 (1970).

CHAPTER 9

INTERCONVERSION OF PURINE RIBONUCLEOTIDES

I. Introduction and Early History

The conversion to uric acid of dietary adenine and guanine and their respective nucleosides and nucleotides was discovered relatively early in the study of purine metabolism in animals; the pathways involved are discussed in Chapter 10. The recognition that adenine and guanine were interconverted, however, was a relatively late development; the history of this topic has been reviewed by Christman (1) and Magasanik (2).

Brown and co-workers in 1948 (3) fed ^{15}N-adenine to rats and found that both adenine and guanine in the visceral nucleic acids were labeled and that the specific activity of guanine was 60% that of adenine. Partly because of rapid deamination of guanine, especially in rats, similar studies with ^{15}N-guanine did not show a corresponding synthesis of labeled nucleic acid adenine. However, later studies by Abrams and Bentley (4), with slices of erythroid bone marrow from rabbits, demonstrated conversion of guanine to adenine and at least some mutual interconversion of adenine and guanine has been found in most animal cells by the use of sufficiently sensitive techniques.

Studies of the nutritional requirements for purines of certain microorganisms also led to the recognition that adenine and guanine—or their derivatives—could be interconverted. For example, *Tetrahymena geleii* requires guanine for growth, and ^{14}C-guanine was shown to be converted to nucleic acid adenine. Magasanik also isolated numerous mutants of enterobacteria which could not synthesize purines *de novo*, but which could grow on adenine, guanine, hypoxanthine, or xanthine. That this class of mutant could respond to any one of these purines indicates that they are capable of interconverting them or their derivatives.

Having established that purines can be interconverted in the course of the incorporation of bases into nucleic acids, it remained to determine whether this was accomplished at the level of the bases themselves, through the ribonucleosides or ribonucleotides, or at more than one level. Early studies of purine metabolism had shown the existence of the following catabolic pathways (see Chapter 10):

Adenosine → inosine
              ↓
Adenine → hypoxanthine → xanthine → uric acid
                            ↑
Guanosine ————————————→ guanine

Most of these reactions are irreversible, and thus they cannot account for the observed interconversion of adenine and guanine.

The findings that inosinate was the end product of the pathway of *de novo* synthesis and that adenylate and guanylate were the first metabolites formed in the course of the incorporation into nucleic acids of adenine and guanine, were most consistent with interconversion at the ribonucleotide level. This conclusion has been confirmed in all subsequent studies with animal cells; some exceptions in bacteria will be mentioned below.

II. Adenylate and Guanylate Synthesis from Inosinate

Studies of the interconversion of purine ribonucleotides began with inosinate. Because radioactive glycine labeled both nucleic acid adenine and guanine, it was apparent that inosinate had to be converted to adenylate and guanylate.

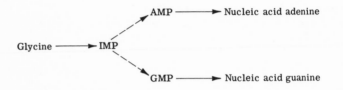

It was also clear that microorganisms living in purine-free environments synthesized their nucleic acid purines *de novo* via inosinate.

Studies with *Corynebacterium diphtheriae* were also informative; it requires for growth either hypoxanthine alone, or a mixture of adenine

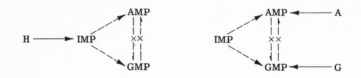

and guanine (5). Thus, this organism cannot interconvert adenine and guanine, but can convert hypoxanthine, presumably as inosinate, to nucleotides of both other bases.

The synthesis of adenylate from inosinate in bone marrow and microbial preparations, and of guanylate from inosinate in pigeon liver, bone marrow, and bacterial extracts were reported between 1955 and 1957 (see references *2, 6–10*). In the course of these studies, two new purine ribonucleotides were identified and shown to be intermediates in these processes.

A. ROLE OF ADENYLOSUCCINATE

Although adenylosuccinate is an intermediate in the synthesis of adenylate, it was first identified as a product of the reaction of adenylate and fumarate. Lieberman (*11*) showed that the synthesis of adenylosuccinate from inosinate and aspartate specifically required GTP and that during the reaction GDP and inorganic phosphate were formed. The requirement for aspartate as nitrogen donor was also specific.

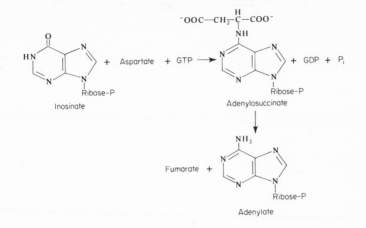

B. ROLE OF XANTHYLATE

Xanthylate was first identified as a product of the NAD^+-dependent oxidation of inosinate; its role as an intermediate in guanylate synthesis had previously been deduced because the accumulation of xanthosine was detected in a guanine-requiring mutant microorganism. Avian and mammalian enzymes which catalyze the amination of xanthylate require glutamine but can utilize ammonia to a small extent; bacterial enzymes require ammonia. Neither the synthesis nor amination of xanthylate is reversible.

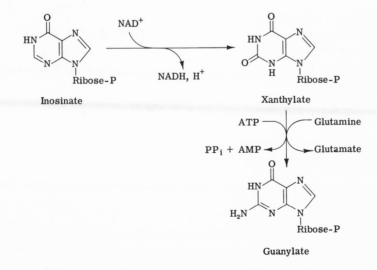

C. NUCLEOTIDE INTERCONVERSION IN BACTERIAL MUTANTS

The above-mentioned reactions of adenylate and guanylate synthesis have been confirmed by the use of bacteria and bacterial mutants which require specific purines for growth.

Some mutants require either xanthine or guanine for growth and have been shown to lack the enzyme responsible for the conversion of inosinate to xanthylate.

Other mutants require guanine specifically and excrete xanthosine; they lack the second enzyme in this pathway.

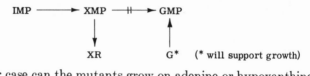

In neither case can the mutants grow on adenine or hypoxanthine, thereby demonstrating the essential character of this pathway of guanylate synthesis.

Among mutants which require adenine for growth, some have been shown to lack the enzyme which makes adenylosuccinate, and others, that which cleaves it to adenylate. The fact that the loss of either of these enzymes results in a specific requirement for adenine proves the essential nature of these steps for the production of adenylate from ribonucleotides of hypoxanthine, xanthine, and guanine.

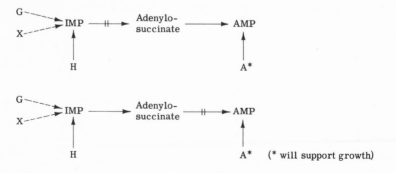

III. Inosinate Synthesis from Guanylate and Adenylate

The conversion of guanine to adenylate must proceed through the steps leading from inosinate to adenylate; otherwise, the adenine-requiring mutants mentioned above which are blocked in these steps should be able to grow on guanine alone.

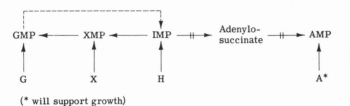

(* will support growth)

Thus, guanine must first be converted to inosinate, and this conversion cannot take place by the reversal of the reactions leading from inosinate to guanylate because these reactions are irreversible. Thus, the existence of another route is indicated (dashed line above).

A. HYDROLYTIC AND REDUCTIVE DEAMINATIONS

These and other observations pointed to the direct formation of inosinate from a derivative of guanine, and an enzyme was discovered which catalyzes the irreversible reductive deamination of guanylate to inosinate.

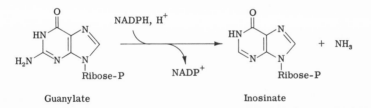

The loss of this enzyme by mutation makes it impossible for the cell to convert guanine to adenylate.

The conversion of adenine to guanylate does not involve the reversal of the steps leading from inosinate to adenylate, for these reactions strongly favor adenylate synthesis. However, mutants blocked in one of these reactions were capable of converting adenine to guanylate and the existence of another route from adenylate to inosinate was indicated (dashed line).

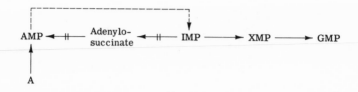

In most animal cells, and in some microorganisms, adenylate is directly converted to inosinate by hydrolytic deamination.

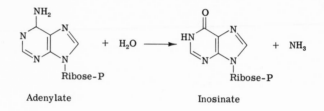

B. ALTERNATIVE PATHWAYS

In some microorganisms the conversion of adenylate to inosinate takes place via the reactions of histidine synthesis from ATP, as described in Chapter 3.

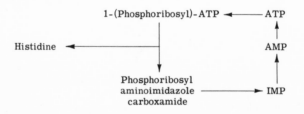

In enteric bacteria which do not make histidine, still other pathways may be operative.

The main reactions of purine ribonucleotide interconversion may be summarized as follows:

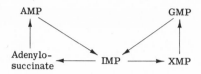

IV. Individual Enzymes

Having established the general routes of purine ribonucleotide interconversion, the six principal enzymes involved will be discussed in more detail.

A. ADENYLOSUCCINATE SYNTHETASE

This enzyme catalyzes the reaction

$$IMP + aspartate + GTP \rightarrow adenylosuccinate + GDP + P_i$$

The substrate specificity of adenylosuccinate synthetase is strict: inosinate cannot be replaced by hypoxanthine, inosine, or phosphoribosyl aminoimidazole carboxamide; L-aspartate cannot be replaced by ammonia or other D- or L-amino acids; and other nucleoside mono-, di-, and triphosphates cannot replace GTP. Divalent cations are required. Bacterial, plant, and animal enzymes have similar specificities and general properties, although Michaelis constants differ somewhat. These are approximately 10^{-4} to 10^{-5} M for GTP and aspartate, and 3×10^{-5} M for inosinate.

Rudolph and Fromm (*12*) have recently made a thorough kinetic study of this enzyme and found the binding of substrates and dissociation of products to be random; however, covalent enzyme–substrate intermediates did not take part in the reaction. They concluded that it is still not possible to decide between the mechanism proposed by Lieberman (*11*), involving a 6-phosphoryl inosinate intermediate, and the simultaneous interaction of the reactants in a concerted transition complex as suggested by Miller and Buchanan (*13*).

B. ADENYLOSUCCINATE LYASE

This enzyme is specific for the substrates of the reverse reaction, adenylate and fumarate, although there is some activity with deoxyadenylate.

$$Adenylosuccinate \leftrightarrows fumarate + AMP$$

The apparent Michaelis constants differ with enzyme source and pH, but are about 5×10^{-5} M for adenylate, 5×10^{-4} M for fumarate, and 10^{-6} to 10^{-5} M for adenylosuccinate. No metal ion is required. Woodward and Braymer (14) found that the *Neurospora* enzyme has a molecular weight of 200,000 and is composed of six to eight subunits. Individual subunits were inactive.

The kinetics and mechanism of adenylosuccinate lyase have been studied thoroughly (15–19). The reaction is primarily ordered, with fumarate dissociating from the enzyme before adenylate; fumarate cannot bind to free enzyme. The succinyl moiety of adenylosuccinate participates in binding, and the dissociation constant of the substrate is eight-fold less than that of adenylate. A sulfhydryl group is at or near the active site and is essential for activity. Reaction of fumarate with adenylate involves *trans* addition of the amino group across the double bond and the removal of fumarate from adenylosuccinate is by an equivalent mechanism. Breakage of the carbon–nitrogen and carbon–hydrogen bonds of the substrates are not the rate-limiting steps of the reaction.

As mentioned in Chapter 7, adenylosuccinate lyase not only functions in purine ribonucleotide interconversion, but also converts phosphoribosyl aminoimidazole succinocarboxamide to phosphoribosyl aminoimidazole carboxamide. Kinetic studies show that these compounds are alternative substrates and products, respectively, with adenylosuccinate and adenylate, and attempts physically to separate the two activities have consistently failed. Genetic studies have shown that both activities are governed by the same locus and that mutant enzymes behave similarly in both reactions (20, 21).

C. ADENYLATE DEAMINASE

This enzyme was first discovered by Schmidt in 1928 in muscle, and the extensive literature regarding it has been reviewed by Lee (22). This enzyme catalyzes the following reaction:

$$AMP + H_2O \rightarrow IMP + NH_3$$

It is specific for 5'-adenylate, but there is some activity with 5'-deoxyadenylate. The Michaelis constant of the crystalline muscle enzyme is 1.4×10^{-3} M, and the optimum pH is 6.0 to 6.5 (23). The molecular weight is about 270,000 (24).

D. INOSINATE DEHYDROGENASE

Bacterial, plant, and animal inosinate dehydrogenases require NAD^+ as the electron acceptor, and either K^+ or NH_4^+, as well as thiols, for maximum activity.

$$IMP + NAD^+ + H_2O \rightarrow XMP + NADH$$

The Michaelis constants are similar (2×10^{-5} M for inosinate; 0.2 to 4×10^{-3} M for NAD$^+$) for enzymes from different sources, with the exception of Sarcoma 180 tumor cells, where that for NAD$^+$ is 5×10^{-5} M (25). Both the *Aerobacter aerogenes* enzyme (26) and that from Sarcoma 180 tumor cells (25) have been shown to catalyze strictly ordered reactions in which inosinate and xanthylate bind to free enzyme. The order of binding of other substrates (including H$_2$O and NAD$^+$) could not be determined. The phosphate of inosinate is important for binding of this substrate; hence the dissociation constants of the 5'-phosphorothioate, 5'-thio, and 5'-amino analogues of inosinate are higher than that of inosinate (27, 28).

E. Guanylate Synthetase

As would be expected for a glutamine amide transfer reaction (Chapter 5), the animal enzyme prefers glutamine as amino group donor, but can use ammonia to a lesser extent.

XMP + ATP + glutamine (or NH$_3$) + H$_2$O → AMP + GMP + PP$_i$ + glutamate

The Michaelis constants for the enzyme from thymus were 5×10^{-4} M (glutamine) and 2×10^{-2} M (NH$_3$ + NH$_4^+$) (4). However, if uncharged NH$_3$ is the true substrate as it is for the bacterial enzyme, then the two amino group donors are equally good substrates.

In studies of the mechanism of guanylate synthetase it was first observed that ^{18}O from the 2-position of xanthylate appeared in the phosphate moiety of adenylate. In the absence of NH$_3$, an intermediate was detected which contained adenylate and xanthylate in equimolar quantities, and its formation was accompanied by synthesis of pyrophosphate. Upon addition of NH$_3$, the adenylate-xanthylate intermediate (whose structure has not been rigorously characterized) was cleaved to adenylate and guanylate.

XMP + ATP → [AMP-XMP]-enzyme + PP$_i$

[AMP-XMP]-enzyme + NH$_3$ → GMP + AMP

In the presence of hydroxylamine, the hydroxylamine analogue of guanylate is readily formed, but fails to dissociate from the enzyme; consequently hydroxylamine is a potent inhibitor   (29).

Enzyme-[AMP-XMP] + NH$_2$OH → Enzyme-[6-hydroxy-2-hydroxylamino-purine] + AMP

F. Guanylate Reductase

Although there is a good deal of evidence for low activities of this enzyme in many animal cells, it has been assayed specifically only in one cell type, the human erythrocyte (30). The enzyme is highly active in *Aerobacter aerogenes*; most studies have employed this enzyme, which is specific for guanylate and NADPH and which requires a sulfhydryl activator.

$$\text{NADPH} + \text{GMP} \rightarrow \text{NADP}^+ + \text{IMP} + \text{NH}_3$$

Michaelis constants of 7.2 μM for guanylate and 13 μM for NADPH were reported by Brox and Hampton (31). An enzyme sulfhydryl appears to be at or near the active site (26). The kinetic mechanism of this reaction is random bi-ter; the two substrates may bind in any order, and the three products may be released in any order.

V. Function and Operation

The reactions of purine ribonucleotide interconversion may be arranged in two cycles which have inosinate as the common intermediate. One may ask exactly what is the function of these two cycles, into which there are so many points of entry; the question may also be asked whether these pathways actually function in a cyclic manner.

There is little evidence regarding the cyclic operation of these reactions. One cycle may function in muscle where the extensive deamination of adenylate accompanying muscle function must be followed by its rapid resynthesis. McFall and Magasanik (32) have suggested that in cultured L cells, guanine is converted to adenine nucleotides for storage and then converted back to guanine nucleotides for incorporation into nucleic acids.

In general, however, these reactions probably act as linear sequences between adenylate, inosinate, or guanylate and the major end products, ATP and GTP.

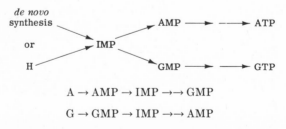

$$\text{A} \rightarrow \text{AMP} \rightarrow \text{IMP} \rightarrow\rightarrow \text{GMP}$$

$$\text{G} \rightarrow \text{GMP} \rightarrow \text{IMP} \rightarrow\rightarrow \text{AMP}$$

Depending on the operation of the *de novo* pathway and on the availability of purine bases, a sequence could start at any of these ribonucleotides and proceed by the most direct route to the two nucleoside triphosphates.

VI. Regulation

The activities of the enzymes of purine ribonucleotide interconversion can be both stimulated and inhibited in a variety of ways, and these potential control mechanisms may function not only to regulate the synthesis of ATP and of GTP, but also to maintain a balance in the relative intracellular concentrations of these two nucleotides.

Increased enzyme activity may be due to increased concentrations of nucleotide substrates, or by other mechanisms. For example, GTP is required as a substrate for adenylosuccinate synthetase, and ATP is required for guanylate synthetase; in addition, ATP activates adenylate deaminase (33).

Most effects of nucleotides on these enzymes, however, are inhibitory. Adenylosuccinate synthetase is inhibited by the product GDP (11), and adenylate, guanylate, and deoxyguanylate are competitive inhibitors with respect to inosinate. Inosinate dehydrogenase is inhibited by guanylate and GDP, which are competitive with respect to inosinate; ATP and GTP are only weak inhibitors (34, 35). The product xanthylate is also a competitive inhibitor with respect to inosinate (25, 26). Guanylate reductase is strongly inhibited by ATP, and weakly by adenylate and inosinate; these effects are reversed by guanylate, although ATP has a distinct binding site (31, 36).

Adenylate deaminase is inhibited by GDP and GTP (37–40) and these inhibitors have mutually antagonistic effects toward the activator ATP (40). The situation is complicated by enzyme activation by Na$^+$ or K$^+$, and in some (but not all) preparations these cations prevent the stimulatory effect of ATP. These phenomena are summarized in Table 9–I.

TABLE 9–I

NUCLEOTIDE INHIBITORS OF ENZYMES OF PURINE
RIBONUCLEOTIDE INTERCONVERSION

Enzyme	Inhibitor
Inosinate dehydrogenase	GMP, GDP, XMP
Guanylate reductase	ATP, AMP, IMP
Adenylosuccinate synthetase	GDP, AMP, GMP, dGMP
Adenylate deaminase	GDP, GTP

The actual physiological role of these stimulatory and inhibitory effects of purine nucleotides on the enzymes of purine ribonucleoside interconversion is very difficult to assess, and little is known about the matter at the present time. Inhibition of inosinate dehydrogenase by metabolites of guanine (36) and inhibition of guanylate reductase by metabolites of adenine (41) appear to occur in some bacteria supplied with these bases.

Repression of the enzymes of purine ribonucleotide interconversion in microorganisms grown in the presence of purines or their derivatives has been demonstrated clearly. Growth of bacteria in the presence of guanine (42–44) or guanosine (45) led to repression of inosinate dehydrogenase; growth on adenosine caused repression of adenylosuccinate synthe ase, and that on adenosine or guanosine caused repression of adenylosuccinate lyase (45).

The supply of nonnucleotide substrates may also be an important control mechanism for these pathways. In Ehrlich ascites tumor cells aspartate concentrations were found to be rate-limiting for adenylate synthesis from hypoxanthine (46), and glutamine levels limited guanylate synthetase activities in these cells and in erythrocytes (47).

Finally, one may ask what are the rate-limiting enzymes for these pathways of purine ribonucleotide interconversion, bearing in mind that this cannot necessarily be determined by their relative maximal activities in cell-free extracts. In Ehrlich ascites tumor cells guanylate reductase is rate-limiting for guanine conversion to adenine nucleotides and adenylosuccinate synthetase is rate-limiting for adenylate synthesis from hypoxanthine. In the absence of added glutamine, guanylate synthetase is rate-limiting for the conversion of both adenine and hypoxanthine to guanine nucleotides, whereas when this amino acid is plentiful, adenylate deaminase and inosinate dehydrogenase, respectively, are limiting (47a).

VII. Effects of Drugs

This subject has been reviewed recently by Balis (48). Three general classes of inhibitors of purine ribonucleotide interconversions may be considered: the 6-thio and 6-chloro purines, amino acid analogues, and psicofuranine.

Inosinate dehydrogenases from bacterial and mammalian sources are inhibited by ribonucleotides of 6-mercaptopurine and 6-thioguanine. Inosinate protects the enzyme against this inhibition, and its analogues are believed to react at the active site of the enzyme. The extents of inhibition by 6-thiopurines and 6-chloropurine are progressive with time, and these inhibitors bind covalently to the enzyme either through a disulfide bond (6-thio compounds) or through a carbon–sulfur bond following displacement of the 6-chloro group (*26, 35, 49*). Similar effects of these analogues on guanylate reductase have also been observed, and the same kind of inhibition mechanism is involved (*31*).

6-Mercaptopurine ribonucleotide and related compounds also inhibit adenylosuccinate synthetase (*28*) and adenylosuccinate lyase (*49–51*).

The glutamine analogue, diazo-oxo-norleucine, and the aspartate analogue, hadacidin (*N*-formyl hydroxyaminoacetic acid), inhibit guanylate synthetase and adenylosuccinate synthetase, respectively. Alanosine (2-amino-3-nitrohydroxylaminopropionic acid) is also an inhibitor of adenylate synthesis from inosinate, but its mechanism of inhibition is not yet clear (*52*).

Psicofuranine (9-D-psicofuranosyl adenine) is an inhibitor of bacterial guanylate synthetase. It binds to a separate site on the enzyme and prevents the synthesis of the adenylate-xanthylate intermediate of this reaction (*29, 53*).

TABLE 9–II

ENZYMES OF PURINE RIBONUCLEOTIDE INTERCONVERSION[a]

Enzyme Commission number	Systematic name	Trivial name
1 1.2.1.14	IMP:NAD oxidoreductase	IMP dehydrogenase
2 6.3.4.1	Xanthosine-5′-phosphate:ammonia ligase (AMP)	GMP synthetase
2 6.3.5.2	Xanthosine-5′-phosphate:L-glutamine amido-ligase (AMP)	GMP synthetase
3 1.6.6.8	Reduced-NADP:GMP oxidoreductase (deaminating)	GMP reductase
4 6.3.4.4	IMP:L-aspartate ligase (GDP)	Adenylosuccinate synthetase
5 4.3.2.2	Adenylosuccinate AMP-lyase	Adenylosuccinate lyase
6 3.5.4.6	AMP aminohydrolase	AMP deaminase

[a] Numbers in the first column refer to numbers in the summary diagram (p. 150).

VIII. Summary

The pathways by which purine ribonucleotides are interconverted are summarized within the total context of purine metabolism in the following scheme. The enzymes involved are listed in Table 9–II.

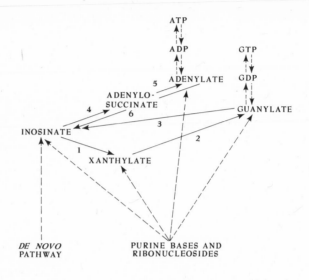

References

1. Christman, A. A., *Physiol. Rev.* **32,** 303 (1952).
2. Magasanik, B., *in* "The Bacteria" (I. C. Gunsalus and R. Y. Stanier, eds.), Vol. 3, p. 295. Academic Press, New York, 1962.
3. Brown, G. B., Roll, P. M., Plentl, A. A., and Cavalieri, L. F., *J. Biol. Chem.* **172,** 469 (1948).
4. Abrams, R., and Bentley, M., *Arch. Biochem. Biophys.* **79,** 91 (1959).
5. Dalby, A., and Holdsworth, E., *J. Gen. Microbiol.* **15,** 335 (1956).
6. Hartman, S. C., and Buchanan, J. M., *Ergeb. Physiol., Biol. Chem. Exp. Pharmakol.* **50,** 75 (1959).
7. Buchanan, J. M., and Hartman, S. C., *Advan. Enzymol.* **21,** 200 (1959).
8. Buchanan, J. M., *in* "The Nucleic Acids" (E. Chargaff and J. N. Davidson, eds.), Vol. 3, p. 304. Academic Press, New York, 1960.
9. Moat, A. G., and Friedman, H., *Bacteriol. Rev.* **24,** 309 (1960).
10. Carter, C. E., *Annu. Rev. Biochem.* **25,** 123 (1956).
11. Lieberman, I., *J. Biol. Chem.* **223,** 327 (1956).
12. Rudolph, F. B., and Fromm, H. J., *J. Biol. Chem.* **244,** 3832 (1969).
13. Miller, R. W., and Buchanan, J. W., *J. Biol. Chem.* **237,** 485 (1962).
14. Woodward, D. O., and Braymer, H. D., *J. Biol. Chem.* **241,** 580 (1966).
15. Cohen, L. H., and Bridger, W. A., *Biochim. Biophys. Acta* **73,** 514 (1963).

16. Cohen, L. H., and Parks, R. E., Jr., *Can. J. Biochem.* **41**, 1495 (1963).
17. Bridger, W. A., and Cohen, L. H., *J. Biol. Chem.* **243**, 644 (1968).
18. Bridger, W. A., and Cohen, L. H., *Can. J. Biochem.* **47**, 665 (1969).
19. Miller, R. W., and Buchanan, J. W., *J. Biol. Chem.* **237**, 491 (1962).
20. Giles, N. H., Partridge, C. W. H., and Nelson, N. J., *Proc. Nat. Acad. Sci. U.S.* **43**, 305 (1957).
21. Gots, J. S., and Gollub, E. G., *Proc. Nat. Acad. Sci. U.S.* **43**, 826 (1957).
22. Lee, Y. P., *in* "The Enzymes" (P. D. Boyer, H. Lardy, and K. Myrbäck, eds.), 2nd rev. ed., Vol. 4, p. 279. Academic Press, New York, 1960.
23. Lee, Y. P., *J. Biol. Chem.* **227**, 999 (1957).
24. Wolfenden, R., Tomozawa, Y., and Bamman, B., *Biochemistry* **7**, 3965 (1968).
25. Anderson, J. H., and Sartorelli, A. C., *J. Biol. Chem.* **243**, 4762 (1968).
26. Brox, L. W., and Hampton, A., *Biochemistry* **7**, 2589 (1968).
27. Hampton, A., Brox, L. W., and Bayer, M., *Biochemistry* **8**, 2303 (1969).
28. Nichol, A. W., Nomura, A., and Hampton, A., *Biochemistry* **6**, 1008 (1967).
29. Fukuyama, T. T., and Donovan, K. L., *J. Biol. Chem.* **243**, 5798 (1968).
30. Hershko, A., Wind, E., Razin, A., and Mager, J., *Biochim. Biophys. Acta* **71**, 609 (1963).
31. Brox, L. W., and Hampton, A., *Biochemistry* **7**, 398 (1968).
32. McFall, E., and Magasanik, B., *J. Biol. Chem.* **235**, 2103 (1960).
33. Smiley, K. L., Jr., Berry, A. J., and Suelter, C. H., *J. Biol. Chem.* **242**, 2502 (1967).
34. Hampton, A., and Nomura, A., *Biochemistry* **6**, 679 (1967).
35. Magasanik, B., Moyed, H. S., and Gehring, L. B., *J. Biol. Chem.* **226**, 339 (1957).
36. Mager, J., and Magasanik, B., *J. Biol. Chem.* **235**, 1474 (1960).
37. Smiley, K. L., Jr., and Seulter, C. H., *J. Biol. Chem.* **242**, 1980 (1967).
38. Setlow, B., and Lowenstein, J. M., *J. Biol. Chem.* **241**, 1244 (1966).
39. Setlow, B., and Lowenstein, J. M., *J. Biol. Chem.* **242**, 607 (1967).
40. Setlow, B., and Lowenstein, J. M., *J. Biol. Chem.* **243**, 3409 (1968).
41. Balis, M. E., Brooke, M. S., Brown, G. B., and Magasanik, B., *J. Biol. Chem.* **219**, 917 (1956).
42. Levin, A. P., *Biochim. Biophys. Acta* **138**, 221 (1967).
43. Levin, A. P., and Magasanik, B., *J. Biol. Chem.* **236**, 184 (1961).
44. Levin, A. P., and Magasanik, B., *J. Biol. Chem.* **236**, 1810 (1961).
45. Nishikawa, H., Momose, H., and Shiio, I., *J. Biochem. (Tokyo)* **62**, 92 (1967).
46. Fontenelle, L. J., and Henderson, J. F., *Biochim. Biophys. Acta* **177**, 88 (1969).
47. Hershko, A., Razin, A., Shoshani, T., and Mager, J., *Biochim. Biophys. Acta* **149**, 59 (1967).
47a. Crabtree, G. W., and Henderson, J. F., *Cancer Res.* **31**, 985 (1971).
48. Balis, M. E., "Antagonists and Nucleic Acids." North-Holland Publ., Amsterdam, 1968.
49. Hampton, A., *J. Biol. Chem.* **238**, 3068 (1963).
50. Hampton, A., *J. Biol. Chem.* **237**, 529 (1962).
51. Atkinson, M. R., Morton, R. K., and Murray, A. W., *Biochem. J.* **92**, 398 (1964).
52. Gale, G. R., and Smith, A. B., *Biochem. Pharmacol.* **17**, 2495 (1968).
53. Kuramitsu, H., and Moyed, H. S., *J. Biol. Chem.* **241**, 1596 (1966).

CATABOLISM OF PURINE NUCLEOTIDES

I. Introduction

Purine nucleotides are broken down by animal cells to fragments which are excreted in order to maintain a relatively constant internal composition in the face of a constant synthesis of these compounds both *de novo* and from dietary constituents. Microorganisms may also catabolize nucleotides, in some cases as sources of energy, carbon, and nitrogen for growth. Although multiple pathways for nucleotide catabolism exist in animals, only certain of these exist in any particular cell. In addition, when alternative pathways do exist within the same cell, it is sometimes very difficult to evaluate the relative importance of each. The overall picture of potential pathways of purine nucleotide catabolism as given in summary in this chapter, therefore, cannot be applied indiscriminately to all cells.

II. Catabolic Enzymes of Uric Acid Synthesis

A number of enzymes which participate in the conversion of purine nucleotides to uric acid are known to occur in nature (1–3), and it is clear that four different kinds of enzymatic reactions are required for this overall process: dephosphorylation, deamination, cleavage of glycosidic bonds, and oxidation.

A. DEPHOSPHORYLATION

Nucleoside di- and triphosphates are dephosphorylated by enzymes of different specificities than those which act on the monophosphates. Nucleoside triphosphates, first of all, may be dephosphorylated both to di- and monophosphates by the many enzymes which use triphosphates functionally as substrates in intermediary metabolism. In addition, they are also acted upon by several enzymes whose physiological role is not clear. For example, inorganic pyrophosphatase in the presence of Zn^{2+}, and apyrase (ATP diphosphohydrolase) degrade both ATP and ADP to adenylate.

Nucleoside diphosphates may be cleaved, as mentioned above, by inorganic pyrophosphatase, and a phosphatase specific for IDP, GDP, and UDP has also been described. Adenylate kinase also may convert ADP to adenylate (plus ATP). Studies of these enzymes have been reviewed by Keilley (4) and Morton (5).

The widely distributed and relatively nonspecific acid and alkaline phosphatases can convert nucleoside monophosphates to nucleosides, regardless of the base or of the position of the phosphate (2', 3', or 5'). Rat liver lysosomes contain an acid nucleotidase which is not specific regarding the position of the phosphate, but which hydrolyzes adenine nucleotides faster than those of other bases (6); it is different from the nonspecific lysosomal sugar phosphatase. Some plants also contain a specific 3'-nucleotidase.

Specific 5'-nucleotidases have been extensively studied (7). There appear to be two general types: one with an alkaline pH optimum which hydrolyzes adenylate most rapidly, and one with an acid pH optimum which dephosphorylates inosinate, xanthylate, and guanylate preferentially; rat liver cells contain both types. Although conflicting reports have been published regarding the intracellular localization of the alkaline 5'-nucleotidase, Bodansky (7) now believes that the rat liver enzyme is confined to the plasma membrane.

5'-Nucleotidases are inhibited to some degree by their nucleoside products. In addition, the cardiac muscle enzyme is inhibited by ATP (8), and the enzyme from sheep brain is inhibited by ATP, UTP, and CTP, but not by GTP (9). The physiological roles of these potential control mechanisms remain to be investigated.

B. DEAMINATION

Specific enzymes exist for the deamination of adenine and guanine and their nucleosides and nucleoside monophosphates, but these enzymes are not uniformly distributed in nature. Most of the deaminases are

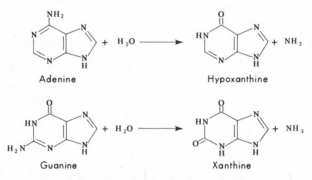

Adenine   Hypoxanthine

Guanine   Xanthine

$$\text{Adenosine} + H_2O \rightarrow \text{inosine} + NH_3$$

$$\text{AMP} + H_2O \rightarrow \text{IMP} + NH_3$$

relatively specific, and all catalyze irreversible reactions. Adenylate deaminase has a catabolic role and is also involved in the cycle of purine nucleotide interconversions; it is known to be regulated by the relative concentrations of ATP (activating) and GTP (inhibiting). There is also a 3'-adenylate deaminase in some animal tissues and an adenylate deaminase in *Aspergillus oryzae* which is not specific regarding the position of the phosphate.

Adenosine deaminase has a very wide distribution and has been studied extensively (see, e.g., reference 10). A large number of adenosine analogues will serve as substrates, including some which contain chloro and other groups at the 6-position. Enzymes from different sources are not identical with respect to Michaelis constants or maximum velocities, and their catalytic mechanisms may differ. Multiple forms of adenosine deaminase may be found even within a single tissue, and the patterns of inheritance of its isozymes have been studied (11). In the heart this enzyme may serve an important regulatory function by inactivating the potent vasopressant agent, adenosine (12).

Adenine deaminase is not found in animal tissues, but is present in some plants and microorganisms.

Guanylate reductase, which deaminates this nucleotide, catalyzes a reductive, rather than hydrolytic, deamination and has been discussed in Chapter 9. Like adenylate deaminase, it has a catabolic role and also functions in purine nucleotide interconversion. A guanosine deaminase has recently been identified in a pseudomonad (13), but it is not known to occur in animal cells.

Guanine deaminase (formerly called guanase) has been well studied; particularly high concentrations are found in rat brain and erythrocytes. It is specific in its substrate requirements, and only guanine and 8-aza-guanine are deaminated effectively. Enzymes from liver, brain, and muscle have different pH optima. Guanase activity is not detectable in rat liver at birth, but its activity increases rapidly thereafter (14); its subcellular distribution changes during the several weeks after birth, as does that of a protein inhibitor of this enzyme (15).

C. Glycosidic Bond Cleavage

In theory, glycosidic bond cleavage may be accomplished either phosphorolytically or hydrolytically and at either the nucleotide or nucleoside level. In most biological systems, the major reaction is the phosphorolytic cleavage of nucleosides.

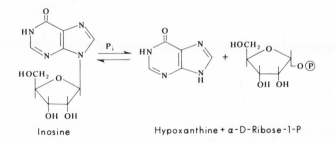

Inosine   Hypoxanthine + α-D-Ribose-1-P

This reaction is reversible, and the equilibrium lies in the direction of nucleoside synthesis (16). Purine nucleoside phosphorylase activities in different cells appear to have different substrate specificities, and it is not clear if there is a single enzyme whose specificity varies or if there are several phosphorylases with different specificities.

Both ribose-1-P and deoxyribose-1-P serve as sugar donors in the synthetic reaction. In addition to the naturally occurring purines, certain purine analogues can also serve as substrates or inhibitors of purine nucleoside phosphorylase. Of these, 8-azaguanine, 6-mercaptopurine, 6-thioguanine, and allopurinol bind most tightly to the human erythrocyte enzyme (17).

Until recently it has generally been believed that adenine was not a substrate of purine nucleoside phosphorylase of animal cells. Zimmerman et al. (18), however, have found that a purified human erythrocyte enzyme synthesizes adenosine at the same maximum rate as it synthesizes inosine, but the Michaelis constant for adenine is much higher than that for hypoxanthine.

Nucleoside hydrolases have been discovered in microorganisms, and these have relatively little specificity for their nucleoside substrates. A

$$BR + H_2O \rightarrow B + \text{ribose}$$

hydrolase for adenylate has also been demonstrated in some bacteria.

D. OXIDATION

Hypoxanthine and xanthine can be oxidized to uric acid by xanthine oxidase (see references 19, 20).

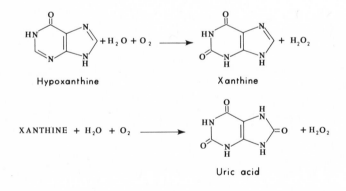

The specificity of this enzyme is low; it will convert adenine to its 8-hydroxy and 2,8-dihydroxy derivatives and will also oxidize a wide variety of nonpurine substrates. Xanthine oxidase contains 2 moles of FAD, 2 moles of $Mo(V)$, and 8 moles of Fe^{3+}; its molecular weight is 200,000–

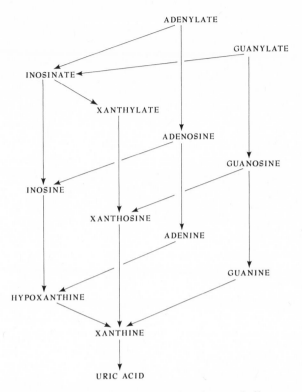

Fɪɢ. 10-1. Catabolic pathways of purine metabolism.

400,000. It has been suggested that hypoxanthine is bound through its 1-, 3-, and 7-nitrogens, and that the xanthine thus formed dissociates from the enzyme and binds through its 3-, 7-, and 9-nitrogens for oxidation to uric acid. In addition to the mammalian enzyme, which uses methylene blue or NAD$^+$ as immediate electron acceptors, avian tissues and some bacteria contain a xanthine dehydrogenase which does not require these compounds for activity. Stirpe and Della Corte (21) have recently suggested that the dehydrogenase is the naturally occurring form in all tissues, but that the oxidase is formed during storage of tissues, or during preparration.

Figure 10-1 draws the most common of these enzyme reactions together in an overall scheme (adopted from Moat and Friedman, 22).

E. Fᴜɴᴄᴛɪᴏɴ ᴀɴᴅ Rᴇɢᴜʟᴀᴛɪᴏɴ

The enzymes mentioned briefly above provide several alternative pathways for the catabolism of purine ribonucleotides to xanthine and for the oxidation of xanthine to uric acid. In addition, the five nucleotides linked in the cycles of nucleotide interconversion provide as many exit points into the catabolic processes as they do entry points into the anabolic processes. In this section the alternative pathways followed by various cells will be outlined, inasmuch as they are known, and some points regarding the regulation of the catabolic pathways will be mentioned.

1. Dephosphorylation

Although several enzymes can in theory convert purine ribonucleotides to ribonucleosides, in only a few cases is it known exactly which is acting. Certain strains of *Bacillus subtilis* accumulate nucleotides extracellularly, but mutants lacking 5'-nucleotidase or alkaline phosphatase, or both, accumulate smaller quantities of nucleotides than cells which have both enzymes. 5'-Nucleotidase appeared to be quantitatively the more important for nucleotide dephosphorylation in these cells (*23*). Baer and Drummond (*12*) compared the rates of dephosphorylation of the 2'-, 3'-, and 5'-phosphates of adenosine by perfused rat heart and concluded that only a specific 5'-nucleotidase was actively in contact with the blood. More evidence for the action of specific, rather than nonspecific phosphatases, are the observations that in one system or another, adenylate, inosinate, xanthylate, or guanylate appears to be the nucleotide most rapidly dephosphorylated.

The nucleoside monophosphate most susceptible to dephosphorylation and the pathway to hypoxanthine or xanthine subsequently followed vary from one cell type to another. Among the patterns that have been observed are the following:

$$AMP \rightarrow AR \rightarrow HR \rightarrow H$$
Heart (*12, 24*)

$$AMP \rightarrow IMP \rightarrow HR \rightarrow H$$
Ehrlich ascites tumor cells (*25*), muscle (*26*), kidney (*27*)

$$AMP \rightarrow IMP \rightarrow XMP \rightarrow XR \rightarrow X$$
Erythrocytes (*28*)

$$GMP \rightarrow IMP \rightarrow XMP \rightarrow XR \rightarrow X$$
$$\downarrow$$
$$HR \rightarrow H$$
Erythrocytes (*28*)

2. Adenylate Deaminase

Studies by Burger and Lowenstein (*29*) illustrate some of the complexities of these alternative pathways of nucleotide catabolism and their control. Pathways of deamination and dephosphorylation of adenylate

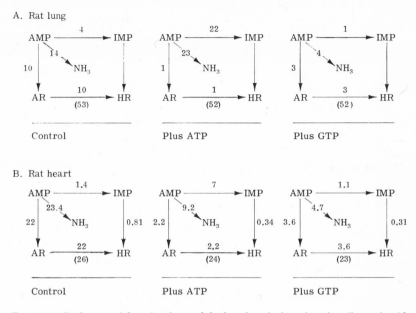

FIG. 10-2. Pathways of deamination and dephosphorylation of purine ribonucleotides. All data given are in mμmoles/mg of protein per minute. From (29). Reproduced with permission.

were measured in soluble fractions of rat lung and heart in the presence and absence of ATP and GTP, the "allosteric" activator and inhibitor, respectively, of adenylate deaminase. It is seen from Fig. 10-2 that in lung extracts, addition of ATP markedly increased total ammonia formation from adenylate by a stimulation of adenylate deaminase that was greater than its inhibition of adenylate dephosphorylation. GTP inhibited adenylate deaminase and its dephosphorylation, and hence the overall rate of ammonia formation was also decreased. Adenosine deaminase deaminated all of the adenosine formed, and the rate of adenosine synthesis was therefore rate-limiting for inosine production.

In rat heart the control rate of adenylate deaminase activity was lower, and that of adenylate dephosphorylation higher than in lung. Most of the ammonia formed from adenylate was therefore due to adenosine deaminase activity. ATP stimulated adenylate deaminase to almost the same relative degree in heart as in lung, but due to a marked inhibition of dephosphorylation the total amount of ammonia formed was less in heart. These data also raise questions concerning the identity and substrate specificities of the enzyme(s) that dephosphorylate adenylate and inosinate.

This study is complicated by the possibility of adenosine kinase activity, which was not considered by the authors, and is undoubtedly oversimplified compared to the situation in intact tissues; however, it does illustrate the problems which must be considered in attempting to evaluate the alternative pathways of nucleotide catabolism.

3. Xanthine Oxidase

Although there is no doubt that xanthine oxidase does function *in vivo* in uric acid synthesis, the regulation of its activity presents some difficult problems, as Knox and Greengard (*30*) have pointed out. Two-thirds of the total xanthine oxidase activity of the rat is in liver, where it is present as a soluble protein. As might be expected from its cofactor composition, diets deficient in riboflavin, molybdenum, or iron lead to decreased enzyme activity. Enzyme activity decreases by 85–90% on protein-free diets, and within this range its activity is generally proportional to the protein content of the diet. Interestingly, allantoin (the urinary end product of purine metabolism in rats) excretion did not decrease proportionally with enzyme activity, which infers that xanthine oxidase is normally present in vast excess. High levels of dietary protein, however, do not increase its activity above normal.

These results are apparently irreconcilable with other studies that show that when xanthine oxidase activity does increase above normal (e.g., in tumor-bearing rats, vitamin E deficiency, and some virus infections) there is always a proportionate increase in allantoin excretion. This increase may be as much as tenfold and is too great to be accounted for by increased nucleic acid turnover under these conditions. Xanthine oxidase activity also increases upon refeeding after starvation, and this has been shown to be due to new RNA and protein synthesis (*31*). These results raise questions concerning intracellular compartmentation, enzyme latentiation, etc., that have yet to be answered.

Allopurinol (4-oxopyrazolo[3,4-*d*]pyrimidine) is a potent inhibitor of

Allopurinol

xanthine oxidase and hence, of uric acid synthesis. It is used clinically in the treatment of gout (see below) and other conditions in which rapid rates of purine catabolism might lead to deleterious precipitation of uric acid in the kidneys.

Hochstein and Utley (*32*) have suggested the possibility that at least

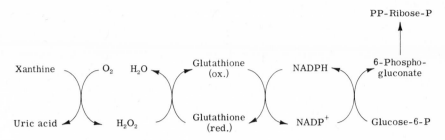

FIG. 10-3. Possible relationship between purine catabolism and purine nucleotide synthesis.

in some cells xanthine oxidase may have a positive feedback influence on purine nucleotide synthesis. The scheme they imply is given in Fig. 10-3 and links xanthine oxidase activity to increased ribose phosphate synthesis via glucose-6-P dehydrogenase and 6-phosphogluconate dehydrogenase activity, through peroxidase and glutathione reductase.

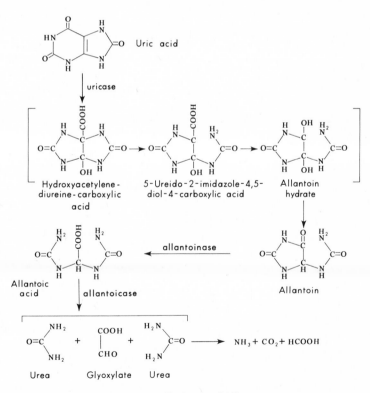

FIG. 10-4. Purine catabolism.

III. Breakdown of Uric Acid

Although uric acid is the end product of purine metabolism in man and higher apes and the end product of nitrogen metabolism as a whole in some other animals, many higher organisms convert it to nonpurine derivatives.

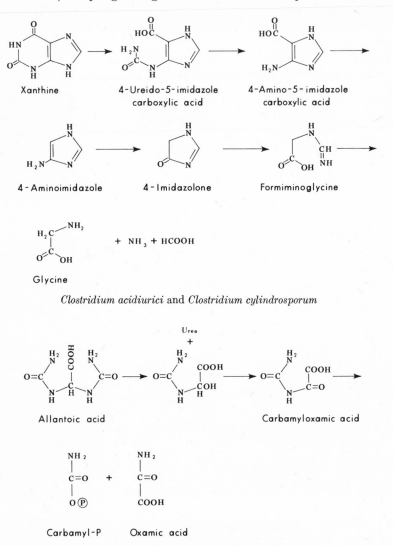

Clostridium acidiurici and *Clostridium cylindrosporum*

Streptococcus allantoicus

FIG. 10-5. Unique microbial pathways of purine catabolism.

In addition, various microorganisms can ferment purines as sources of energy, carbon, and nitrogen. Figure 10-4 shows how uric acid can be broken down all the way to carbon dioxide and ammonia by animals and by such microorganisms as *Aspergillus nidulans*, certain pseudomonads, and *Penicillium chrysogenum* (see reference *33*). Ureidoglycollate is believed to be a product of the allantoicase reaction in *A. nidulans*; this is then converted to urea and glyoxylate.

Several intermediates are either known or postulated to exist in the uricase reaction (Fig. 10-4), whose final product is allantoin. Studies with labeled uric acid have shown that C-6 is eliminated in this reaction as CO_2. Because a label (^{15}N) at N-1 is found equally in the ureide and hydantoin portions of allantoin, the existence of a symmetrical intermediate such as hydroxyacetylene-diureine-carboxylic acid has been assumed.

Little is known about the enzymes allantoinase and allantoicase.

Figure 10-5 shows several unique microbial pathways of purine catabolism (see reference 34).

The breakdown of xanthine by *Clostridium acidiurici* and *C. cylindrosporum* is of interest not only because this purine can be the sole energy source of these microorganisms, but also because many of the intermediates of this pathway resemble dephosphoribosylated intermediates of the pathway of purine biosynthesis *de novo*.

IV. Comparative Aspects of Purine Catabolism

It has already been stated that uric acid is the principal end product of purine metabolism in man and the higher primates, while allantoin is formed in other mammals. Other groups in the animal kindgom also have characteristic end products: allantoic acid, urea, or ammonia. This pattern is, of course, due to the phylogenetic distribution of the various enzymes involved in uric acid catabolism, and a very generalized picture of this distribution is shown in Table 10–I (*35*).

The specific excretion of hypoxanthine by leeches and of guanine by spiders presents interesting patterns of enzyme specialization. Obviously, a number of catabolic enzymes present in most other organisms are missing, but detailed enzyme studies of these creatures have not yet been made. A lack of guanine deaminase leads to a high excretion of this purine by pigs, although allantoin is also formed.

A. Uricotelism

Most of the animals listed in Table 10-I are ureotelic or ammoniotelic; i.e., urea or ammonia is the chief end product of *protein* metabolism. Birds,

TABLE 10–I

END PRODUCTS OF PURINE METABOLISM
IN ANIMALS[a]

Major excretory product	Representative animals
Uric acid	Man and higher primates
↓ uricase	Birds and reptiles
	Insects
Allantoin	Other mammals
↓ allantoinase	Diptera
	Gasteropods
Allantoic acid	
↓ allantoicase	
Urea	Fish
	Amphibia
↓ urease	Lamellibranchs (fresh water)
Ammonia	Lamellibranchs (marine)
	Crustacea
	Sipunculids
Guanine	Spiders
	(high in pigs, snails)
Hypoxanthine	Leeches
	Anodonta

[a] From (35). Reproduced with permission.

terrestrial reptiles, and many insects, however, form uric acid as the major end product of both purine and protein metabolism, and hence are termed uricotelic. It is considered that the excretion of uric acid is an advantage to these uricotelic groups, which live in environments in which water is scarce. It precipitates due to low solubility, and hence exerts no osmotic effect. Urinary water can be reabsorbed and a solid or semisolid excreta formed. The same process occurs in the water-impermeable (cleidoic) eggs of these animals where it is also rendered osmotically inactive by precipitation.

The excretion of uric acid in uricotelic animals is an interesting process. The chicken, for example, forms by glomerular filtration and tubular secretion a ureteral urine which is 21% uric acid, a concentration about 3000 times the blood level. As the urine passes to the cloaca, most of the uric acid precipitates, and the osmotically equivalent water is then reabsorbed by rectal or anal glands. The semisolid uric acid paste is then defecated with the intestinal residues. Other organisms have special glands in which precipitated uric acid is stored for periodic or occasional excretion. The nephridia of terrestrial gastropods, for example, may contain 150 to 1000 mg of uric acid per gram of dry weight, depending on the stage of their 2 to 16 week excretion cycle (36). Males of six species of cockroach likewise store uric acid and excrete it only during copulation (37).

Many fish produce urea as the major end product of protein metabolism and some synthesize this compound via the well-known ornithine-urea cycle. Other species, however, convert amino acids first to uric acid, and this purine is then converted to urea by the processes described above (38).

B. The Dalmatian Coach Hound

It has been known for many years that the dalmatian is the only known mammal besides the higher primates which excretes uric acid as the end product of its purine metabolism. Uric acid is excreted in the urine despite the presence of normal uricase activity in its liver, and this peculiarity requires genetic homozygosity for expression.

This phenomenon is believed to be due to excretion by the kidneys of the uric acid passing in the plasma from extrahepatic tissues to the liver where uricase is located. Uric acid is completely filtered at the glomerulus and both actively reabsorbed and actively secreted in the proximal tubule (39, 40). In the dalmatian the reabsorptive process is deficient, and active secretion leads to the urinary excretion of uric acid.

Uric acid excretion in other animals is affected by a variety of drugs which act on these active reabsorptive or secretory systems. For example, low doses of probenecid and sulfinpyrazone block the secretory system and

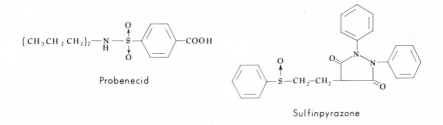

Probenecid

Sulfinpyrazone

hence cause increased blood levels of uric acid. In contrast, higher doses inhibit reabsorption from the kidney tubules, and blood uric acid concentrations decline. These drugs are sometimes useful in gout and other hyper-uricemic conditions.

Pyrazinamide is an antitubercular drug which as a side effect blocks the

Pyrazinamide

kidney urate secretory system, with hyperuricemia as a result.

C. GOUT

Except in the case of uricotelic animals, the evolutionary advantage of the conversion of uric acid to ammonia, urea, allantoic acid, or allantoin, is not always clear. The disadvantage of the lack of uricase in man and the higher primates, however, is that the insolubility of uric acid may lead to its deposition in tissues, thereby causing gout. The formation of uric acid deposits in tissues is preceded by and associated with hyperuricemia, the cause of which has been both much studied and hotly debated. It seems

TABLE 10-II

URINARY PURINES IN MAN[a]

Compound	Daily excretion (mg/day)
Hypoxanthine	9.7
Xanthine	6.1
7-Methylguanine	6.5
8-Hydroxy-7-methylguanine	3.6
Adenine	1.4
6-Succinoaminopurine	1.0
1-Methylguanine	0.6
N^2-Methylguanine	0.5
1-Methylhypoxanthine	0.4
Guanine	0.4
Allantoin	20
Uric acid excretion in: urine	400
saliva	40
gastric juice	8
bile	4

[a] See (42).

likely that there is more than one cause of hyperuricemia, and therefore more than one cause of gout. One type is characterized by the synthesis of purines *de novo* at faster than normal rates, due to increased PP-ribose-P availability or perhaps because the pathway is under less stringent feedback inhibition. Other causes of hyperuricemia may be faulty renal function or altered glutamine metabolism (*41*).

TABLE 10–III

ENZYMES OF PURINE CATABOLISM

Enzyme Commission number	Systematic name	Trivial name
3.6.1.11	Polyphosphate polyphospho-hydrolase	Endopolyphosphatase
3.6.1.5	ATP diphosphohydrolase	Apyrase
3.6.1.6	Nucleosidediphosphate phospho-hydrolase	Nucleosidediphosphatase
2.7.4.3	ATP:AMP phosphotransferase	Adenylate kinase
3.1.3.2	Orthophosphoric monoester phosphohydrolase	Acid phosphatase
3.1.3.1	Orthophosphoric monoester phosphohydrolase	Alkaline phosphatase
3.1.3.5	5'-Ribonucleotide phospho-hydrolase	5'-Nucleotidase
3.1.3.6	3'-Ribonucleotide phospho-hydrolase	3'-Nucleotidase
3.6.1.1	Pyrophosphate phosphohydrolase	Inorganic pyrophosphatase
3.5.4.6	AMP aminohydrolase	AMP deaminase
3.5.4.4	Adenosine aminohydrolase	Adenosine deaminase
3.5.4.2	Adenine aminohydrolase	Adenine deaminase
1.6.6.8	Reduced NADP:GMP oxido-reductase (deaminating)	GMP reductase
3.5.4.3	Guanine aminohydrolase	Guanine deaminase Guanosine deaminase
2.4.2.15	Guanosine:orthophosphate ribosyltransferase	Guanosine phosphorylase
2.4.2.5	Nucleoside:purine (pyrimidine) ribosyltransferase	Nucleoside ribosyl-transferase
2.4.2.1	Purine-nucleoside:orthophos-phate ribosyltransferase	Purine nucleoside phosphorylase
3.2.2.1	*N*-Ribosyl-purine ribohydrolase	Nucleosidase
3.2.2.2	Inosine ribohydrolase	Inosinase
1.3.2.3	Xanthine:oxygen oxidoreductase	Xanthine oxidase
1.7.3.3	Urate:oxygen oxidoreductase	Urate oxidase
3.5.2.5	Allantoin amidohydrolase	Allantoinase
3.5.3.4	Allantoate amidinohydrolase	Allantoicase

V. Urinary Purines Other than Uric Acid

Table 10–II gives the results of one study of urinary excretion in man of allantoin and of purines other than uric acid (42). In addition, 1-methyladenine, N^6-methyladenine, N^2-dimethylguanine, N^6-methyladenosine, adenosine, and 1-methylguanosine have also been reported to be in human urine (43). The methylated purines are believed to be derived mainly from the catabolism of transfer RNA.

Only 60 to 80% of the uric acid synthesized in the human body is excreted in the urine, the remainder being secreted into saliva, gastric juice, and bile. Some of the uric acid secreted into the intestinal tract is broken down to allantoin by intestinal microorganisms.

VI. Summary

The major pathways of purine ribonucleotide catabolism are summarized below. The enzymes involved are listed in Table 10–III.

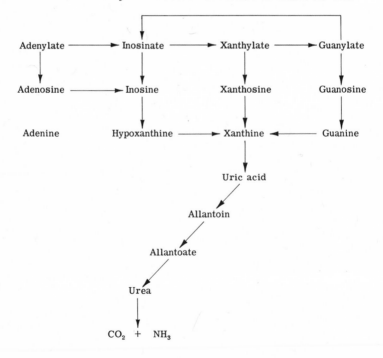

References

1. Schulman, M. P., *Metab. Pathways* **2**, 389 (1961).
2. Potter, V. R., "Nucleic Acid Outlines," p. 217. Burgess Co., Minneapolis, Minnesota, 1960.
3. Balis, M. E., *Advan. Clin. Chem.* **10**, 157 (1967).
4. Kielley, W. W., *in* "The Enzymes" (P. D. Boyer, H. Lardy, and K. Myrbäck, eds.), 2nd rev. ed., Vol. 5, p. 149. Academic Press, New York, 1961.
5. Morton, R. K., *Compr. Biochem.* **16**, 55 (1965).
6. Arsenis, C., and Touster, O., *J. Biol. Chem.* **243**, 5702 (1968).
7. Bodansky, O., and Schwartz, M. K., *Advan. Clin. Chem.* **11**, 277 (1968).
8. Baer, H. P., Drummond, G. I., and Duncan, E. L., *Mol. Pharmacol.* **2**, 67 (1966).
9. Ipata, P. L., *Biochemistry* **7**, 507 (1968).
10. Baer, H. P., Drummond, G. I., and Gillis, J., *Arch. Biochem. Biophys.* **123**, 172 (1968).
11. Spencer, N., Hopkinson, D. A., and Harris, A., *Ann. Hum. Genet.* **32**, 9 (1968).
12. Baer, H. P., and Drummond, G. I., *Proc. Soc. Exp. Biol. Med.* **127**, 33 (1968).
13. Ishida, Y., Sherafuji, H., Kida, M., and Yoneda, M., *Agr. Biol. Chem.* **33**, 384 (1969).
14. Silverstein, E., Zaklynsky, O., and Horn, T., *Proc. Soc. Exp. Biol. Med.* **132**, 91 (1969).
15. Kumar, S., *Arch. Biochem. Biophys.* **130**, 692 (1969).
16. Friedkin, M., and Kalckar, H., *in* "The Enzymes" (P. D. Boyer, H. Lardy, and K. Myrbäck, eds.), 2nd rev. ed., Vol. 5, p. 237. Academic Press, New York, 1961.
17. Krenitsky, T. A., Elion, G. B., Henderson, A. M., and Hitchings, G. H., *J. Biol. Chem.* **243**, 2876 (1968).
18. Zimmerman, M., Gersten, N., and Miech, R. P., *Proc. Amer. Ass. Cancer Res.* **11**, 87 (1970).
19. De Renzo, E. C., *Advan. Enzymol.* **17**, 293 (1956).
20. Bray, R. C., *in* "The Enzymes" (P. D. Boyer, H. Lardy, and K. Myrbäck, eds.), 2nd rev. ed., Vol. 7, p. 533. Academic Press, New York, 1963.
21. Stirpe, F., and Della Corte, E., *J. Biol. Chem.* **244**, 3855 (1969).
22. Moat, A., and Friedman, H., *Bacteriol. Rev.* **24**, 309 (1960).
23. Momose, H., Nishikawa, H., and Katsuya, N., *J. Gen. Appl. Microbiol.* **10**, 343 (1964).
24. Gerlach, E., Deuticke, B., and Dreisbach, R. H., *Naturwissenschaften* **50**, 228 (1963).
25. Overgaard-Hansen, K., *Biochim. Biophys. Acta* **104**, 330 (1965).
26. Lee, C. A., and Newbold, R. P., *Biochim. Biophys. Acta* **72**, 349 (1963).
27. Gerlach, E., Dreisbach, R. H., and Deuticke, B., *Naturwissenschaften* **49**, 87 (1962).
28. Hershko, A., Razin, A., Shoshani, T., and Mager, J., *Biochim. Biophys. Acta* **149**, 59 (1967).
29. Burger, R., and Lowenstein, J. M., *J. Biol. Chem.* **29**, 5281 (1967).
30. Knox, W. E., and Greengard, O., *Advan. Enzyme Regul.* **3**, 247 (1963).
31. Rowe, P. B., and Wyngaarden, J. B., *J. Biol. Chem.* **241**, 5571 (1966).
32. Hochstein, P., and Utley, H., *Mol. Pharmacol.* **4**, 572 (1968).
33. Allam, A., and Elzainy, T., *J. Gen. Microbiol.* **56**, 293 (1969).
34. Barker, H. A., *in* "The Bacteria" (I. C. Gunsalus and R. Y. Stanier, eds.), Vol. 3, p. 181. Academic Press, New York, 1962.
35. Florkin, M., and Duchateau, G., *Arch. Int. Physiol.* **53**, 267 (1943).

36. Needham, J., *Biochem. J.* **29,** 238 (1935).
37. Roth, L. M., and Dateo, G. P., Jr., *Science* **146,** 782 (1964).
38. Forster, R. P., and Goldstein, L., *in* "Fish Physiology" (W. S. Hoar and D. J. Randall, eds.), Vol. 1, p. 313. Academic Press, New York, 1969.
39. Zins, G. R., and Weiner, I. M., *Amer. J. Physiol.* **215,** 411 (1968).
40. Mudge, G. H., Cucchi, J., Platts, M., O'Connell, J. M. B., and Berndt, W. O., *Amer. J. Physiol.* **215,** 404 (1968).
41. Henderson, J. F., *Clin. Biochem.* **2,** 241 (1969).
42. Weissmann, B., Bromberg, P. A., and Gutman, A. B., *J. Biol. Chem.* **224,** 407 (1957).
43. Fink, K., and Adams, W. S., *Arch. Biochem. Biophys.* **126,** 27 (1968).

PART III

PYRIMIDINE RIBONUCLEOTIDE METABOLISM

Pyrimidine ribonucleotides, like those of purines, may be synthesized *de novo* from amino acids and other small molecules (Chapter 11). Preformed pyrimidine bases and their ribonucleoside derivatives, derived from the diet of animals or found in the environment of cells, may be converted to ribonucleotides via nucleoside phosphorylases and nucleoside kinases. In some cells a more direct pyrimidine phosphoribosyltransferase pathway has also been recognized (Chapter 12). Ribonucleotides are catabolized by dephosphorylation, deamination, and cleavage of the glycosidic bond, to uracil. Uracil may be either oxidatively or reductively cleaved, depending on the organism involved, and can be converted to CO_2 and NH_3 (Chapter 13).

The general features of pyrimidine ribonucleotide metabolism are outlined here, and will be discussed in detail in Chapters 11–13.

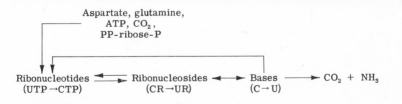

CHAPTER 11

THE *DE NOVO* PATHWAY OF PYRIMIDINE NUCLEOTIDE BIOSYNTHESIS

I. Introduction: Historical Development and Precursors of the Pyrimidine Ring Atoms

Animals do not have a dietary requirement for pyrimidines, and many microorganisms do not require exogenous pyrimidines, because they have the ability to synthesize pyrimidine nucleotides from simple, acyclic precursors. In early investigations of pyrimidine biosynthesis, the analysis of the nucleic acids afforded the only approach available to the determination of cellular pyrimidines, and so incorporation of precursors into the nucleic acid pyrimidines was followed. However, in 1952–1954 as one of the early results of the application of a new nucleotide technology (see Chapter 1), the uridine and cytidine phosphates were recognized both as free solutes in tissues and as metabolic intermediates between precursor molecules and the nucleic acid pyrimidines (*1, 2*). At this time it became apparent that uridine monophosphate was the first nucleotide product containing a nucleic acid base to emerge from the process of pyrimidine synthesis *de novo*.

How were the simple precursors of the pyrimidine ring recognized? In 1943, Barnes and Schoenheimer showed that heavy nitrogen from ^{15}N-ammonium citrate was incorporated into animal polynucleotides. It was apparent as early as 1944 that orotate (6-carboxyuracil) was involved in pyrimidine biosynthesis, because this compound would satisfy the pyrimidine requirement for some bacteria [see review (*3*)]. Between 1949 and 1952 it became apparent that orotate was probably an intermediate in pyrimidine biosynthesis because this compound was utilized efficiently in mammalian tissues as a precursor of both DNA and RNA pyrimidines. However, the possibility remained that orotate per se was not an actual intermediate, but rather was easily converted to one.

In the early 1950s, the small molecule precursors of the pyrimidine ring atoms were identified, principally by Lagerkvist and his group [see review (*4*)]:

(a) In the rat and in Ehrlich ascites tumor cells ammonia was shown to be a major contributor to N-3 of nucleic acid uracil, being incorporated into that position at three or four times the rate of incorporation into N-1.

(b) Carbon dioxide was found to be a specific precursor for the ureide carbon (C-2) of nucleic acid pyrimidines. [This finding has been reaffirmed in more recent work which demonstrated the rapid and very specific incorporation of ^{14}C-bicarbonate into C-2 of the soluble uridine nucleotides in Ehrlich ascites tumor cells (*5, 6*).]

(c) Experiments with labeled aspartate showed that carbon atoms 2, 3, and 4 provided the carbon skeleton of nucleic acid uracil.

(d) The methyl group of nucleic acid thymine was recognized as being derived from a "one-carbon" derivative; formate and carbon-3 of serine were incorporated into this group.

To determine whether orotate was actually an intermediate compound in the process of pyrimidine biosynthesis or, alternatively, entered the sequence by a side reaction, Reichard and Lagerkvist (*7*) undertook "trapping" experiments using liver slices; during *in vitro* incubation the slices incorporated labeled precursors such as ammonia, bicarbonate, or aspartate into RNA pyrimidines. When the slices were incubated with such labeled precursors in the presence of extracellular orotate, isotope was "trapped" in the latter, evidently through equilibration with endogenously formed, labeled orotate. In this way, orotate was recognized as an intermediate in pyrimidine synthesis. Orotate, with isotopic markers introduced in this way from labeled precursors, was degraded by a procedure which allowed the separate isolation of the individual atoms of the ring. By this means, small molecules were recognized as the precursors of particular atoms of the orotate ring, as shown in the following diagram:

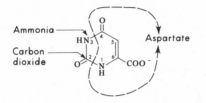

Previously, bacterial nutrition studies had indicated that carbamyl aspartate ("ureidosuccinate" in the earlier literature) might also be an intermediate in pyrimidine synthesis; this compound was also found to be incorporated into orotate by the "trapping" experiments.

Finally, as mentioned above, with the advent of a technology for manipulation of nucleotides, the free uridine nucleotides were recognized as intermediates in the incorporation of orotate into the nucleic acid pyrimidines (2), and it became apparent that the first product of this pathway to possess a nucleic acid base was uridylate.

II. The Orotate Pathway of Uridylate Synthesis

The sequence of reactions which culminates in the formation of uridylate is very widely distributed in nature and is referred to as the "orotate pathway," or as the *de novo* route of pyrimidine biosynthesis. This pathway is summarized in the following diagram; segments of it will be discussed next in the historical sequence of their discovery and development, which is given by the order A, B, C.

Ammonia,   C carbamyl A   B
carbon dioxide, $\longrightarrow$ aspartate $\longrightarrow$ orotate $\longrightarrow$ uridylate
aspartate

A. THE CONVERSION OF CARBAMYL ASPARTATE TO OROTATE

Historically, carbamyl aspartate was recognized as a likely intermediate in pyrimidine biosynthesis because (a) this compound is an assembly of two elementary precursors of the pyrimidine ring, (b) carbamyl aspartate would satisfy the nutritional requirement of *L. bulgaricus* 09 for orotate, and (c) labeled carbamyl aspartate was incorporated into ribonucleic acid pyrimidines in *L. bulgaricus* and, as well, served as an orotate precursor in liver slice "trapping" experiments such as those mentioned above.

The steps in the conversion of carbamyl aspartate to orotate became apparent in work by Lieberman and Kornberg (8), who studied the reverse process, the degradation of orotic acid by an orotate-fermenting bacterium, *Zymobacterium oroticum*. Intact cells of this organism degraded orotate to NH_3, CO_2, acetic acid, and dicarboxylic acids; however, in broken cell preparations, the degradation of orotate did not proceed to that extent and two intermediates were isolated, dihydroorotate and carbamyl aspartate:

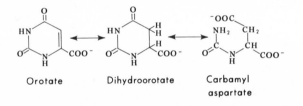

Orotate   Dihydroorotate   Carbamyl aspartate

These degradative reactions were found to be reversible and extracts of *Z. oroticum* readily converted carbamyl aspartate to orotate when incubation conditions provided NAD^+, or catalyzed the reverse sequence under reducing conditions.

1. Dihydroorotase

This enzyme catalyzes the reversible cyclization of the L-isomer of carbamyl aspartate:

$$N\text{-Carbamyl L-aspartate} \longleftrightarrow \text{dihydroorotate} + H_2O$$

Dihydroorotase has been demonstrated in various animal tissues, including the Ehrlich ascites carcinoma (9) and human erythrocytes (10). The latter occurrence is surprising because human erythrocytes do not have the capacity for uridylate synthesis, either by the *de novo* route or from uridine; uridine nucleotides are very minor constituents of the watery portion of these cells.

2. Dihydroorotate Dehydrogenase

Lieberman and Kornberg (11) established the dihydroorotate de-hydrogenase reaction in Z. oroticum as

$$\text{Dihydroorotate} \underset{\text{NADH+H}^+}{\overset{\text{NAD}^+}{\rightleftharpoons}} \text{orotate}$$

The enzyme from this source has since been crystalized and extensively studied [for example, see (12)]. The dehydrogenase is a flavoprotein containing FMN, FAD, and non-heme iron; the reduced flavin in the purified enzyme can be oxidized directly by O_2 and also by NAD^+. The enzyme is specific for orotate and will not react with uracil, cytosine, thymine, or 5-methylcytosine. A related dehydrogenase has also been demonstrated in other bacterial species and in animal tissues.

B.   THE CONVERSION OF OROTATE TO URIDYLATE

With in vivo experiments, Hurlbert and Potter (1) first showed that in rat liver, uridine nucleotides were intermediates in the conversion of orotate to nucleic acid pyrimidines; the first of the three uridine phosphates to become labeled in this process was the monophosphate, uridylate (UMP) (13). The synthesis of uridylate from orotate takes place in two steps: (a) the condensation of orotate with PP-ribose-P to form orotidylate (orotidine 5′-monophosphate, or OMP), and (b) decarboxylation of orotidylate.

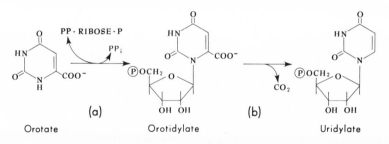

Orotate   Orotidylate   Uridylate

1. Orotate Phosphoribosyltransferase

By fractionating yeast extracts, preparations of the orotate phos-phoribosyltransferase were obtained which were free of the decarboxylase activity (14). When incubated with PP-ribose-P and orotate, these preparations formed orotidylate; this product was identical with orotidylate prepared by the enzymatic phosphorylation of orotidine. The stoichiometry of the reaction was also established.

The yeast orotate phosphoribosyltransferase reaction is unusual in that it is readily reversible, in contrast to the subsequently discovered purine and uracil phosphoribosyltransferase reactions (see Chapters 5 and 12). The formation of orotidylate and the pyrophosphorolysis of orotidylate readily proceed to equilibrium; the equilibrium constant for the forward reaction is about 0.1 (*14*). The reversibility of this reaction was the basis for the earlier name of the enzyme, orotidylate pyrophosphorylase.

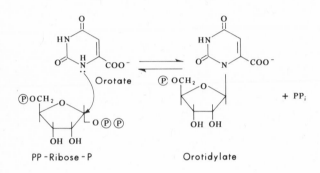

The reaction mechanism is believed to be a nucleophilic attack on the glycosyl carbon atom of PP-ribose-P by an unshared electron pair on N-1 of the pyrimidine ring; displacement of inorganic pyrophosphate results and an *N*-glycosidic linkage is formed in the β-configuration. Thus, an inversion in the configuration of the substituent at C-1 of the ribosyl group occurs during formation of the pyrimidine *N*-glycosidic bond, as in the purine phosphoribosyltransferase reactions (see Chapter 8). Because the yeast enzyme does not accept uracil or dihydroorotate as substrates, this enzyme activity in general has been regarded as specific for orotate. Orotate phosphoribosyltransferase activity has been demonstrated many times in animal tissues as a part of the widely distributed *de novo* pathway of pyrimidine synthesis.

A pyrimidine phosphoribosyltransferase activity with a broader specificity than the yeast enzyme has been demonstrated in animal tissues. Highly purified preparations from calf thymus (*15*) and beef erythrocytes (*16*) accepted orotate and 5-fluorouracil as substrates. Uracil phosphoribosyltransferase activity has also been demonstrated in extracts from mouse leukemia cells. Fluorouracil is a better substrate for this enzyme than uracil at pH 7.5, possibly because the acid dissociation constant for the analogue (pK_a, 8.15) is higher than that of uracil (pK_a, 9.45) (*17*). This reasoning would suggest that the anionic form of the substrate might be the species required by the enzyme. This enzyme has been implicated in the mechanism of action of 5-fluorouracil, which has an important use in cancer chemotherapy (*18*). The pyrimidine phosphoribosyltransferase activity has been found absent from fluorouracil-resistant experimental tumors. It is not yet clear whether there exists in animal cells a pyrimidine phosphoribosyltransferase of broad specificity which is distinct from that of the orotate phosphoribosyltransferase element of the *de novo* pathway of uridylate synthesis.

2. Orotidylate Decarboxylase

The irreversible decarboxylation step, catalyzed by orotidylate decarboxylase, has been demonstrated in various animal tissues.

$$\text{Orotidylate} \longrightarrow \text{uridylate} + CO_2$$

The orotidylate decarboxylases of rat liver and of yeast are competitively inhibited by uridylate and cytidylate, but not by UDP or UTP (*19*).

The activities of this enzyme and orotate phosphoribosyltransferase have usually been measured together by a spectrophotometric assay which depends upon differences between the absorption spectra of the initial substrate, orotate, and the final product, uridylate. This assay has been applied to extracts from many types of cells and has shown that the orotate pathway is a widespread mechanism for uridylate formation.

C. THE FORMATION OF CARBAMYL ASPARTATE

The biosynthesis of carbamyl aspartate from ammonia, carbon dioxide, and aspartate is a two-step process, involving the intermediate formation of carbamyl phosphate.

1. Historical

After it became apparent that carbamyl aspartate was an intermediate in orotate synthesis, Reichard (*20*) investigated the synthesis of carbamyl aspartate in rat liver preparations and demonstrated its formation from aspartate, carbon dioxide, and ammonia, in the presence of ATP and *N*-acetylglutamate. Previously, Grisolia and Cohen (*21*) had proposed that an "active carbamyl" was involved in citrulline synthesis in mammalian liver preparations:

$$\text{Ornithine} + HCO_3^- + NH_4^+ + ATP \xrightarrow[\text{"active carbamyl"}]{\text{via}} \text{citrulline}$$

It was recognized that the "active carbamyl" was probably also involved in carbamyl aspartate formation. Jones *et al.* (*22*) identified the "active carbamyl" as carbamyl phosphate:

$$H_2N-\overset{\overset{\displaystyle O}{\|}}{C}-O-\overset{\overset{\displaystyle O}{\|}}{\underset{\underset{\displaystyle O^-}{|}}{P}}-O^-$$

Carbamyl phosphate

Reichard found that carbamyl aspartate synthesis would take place in crude liver preparations if carbamyl phosphate and aspartate were provided and showed that the enzyme responsible, aspartate carbamyltransferase, was not the same as that which formed citrulline, but apparently shared carbamyl phosphate with it.

2. Aspartate Carbamyltransferase

The enzymatic formation of carbamyl aspartate is reversible, but the equilibrium is much in favor of carbamyl aspartate:

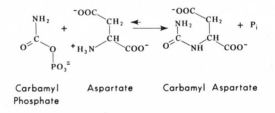

Carbamyl   Aspartate   Carbamyl Aspartate
Phosphate

Aspartate carbamyltransferase activity is widely distributed, having been found in rat tissues, pigeon liver, yeast, and in *E. coli*. Rapidly growing tissues such as tumor and regenerating liver have high concentrations of the enzyme (*23*).

a. Bacterial Aspartate Carbamyltransferase. This enzyme catalyzes the first reaction unique to the *de novo* pathway of pyrimidine nucleotide synthesis and in a number of bacterial species is subject to allosteric inhibition by one of the ultimate products of this reaction, CTP. The *E. coli* enzyme has been very extensively studied because of its regulatory properties (*24*).

The *E. coli* carbamyltransferase has been highly purified and is known to consist of two kinds of subunits, one possessing the catalytic function and the other lacking catalytic activity, but possessing sites for interaction with allosteric effectors (*25*). Upon treatment with *p*-chloromercuribenzoate, the purified native enzyme (mol. wt. 310,000) dissociates into regulatory and catalytic subunits. The native molecule consists of three regulatory subunits and two catalytic subunits. The regulatory subunits each contain two polypeptide chains of 17,000 molecular weight and each subunit has two binding sites for CTP; each catalytic subunit has three polypeptide chains of 33,000 molecular weight and each subunit has three catalytic sites (*26, 27*).

When the catalytic and regulatory subunits are dissociated, the catalytic activity, measured with saturating concentrations of carbamyl phosphate, displays conventional hyperbolic kinetics with respect to aspartate concentration. However, sigmoidal kinetics are shown by the native aspartate carbamyltransferase, in which form the subunits are associated. In the presence of CTP, the sigmoidicity of the rate–aspartate curve is further increased, with the effect that the apparent Michaelis constant for aspartate is increased. ATP competes with CTP for the regulatory site, with the result that the CTP inhibitory effect is antagonized.

The *synthesis* of the *E. coli* aspartate carbamyltransferase is under repressor control (*28*); this is one of the classical examples of metabolite control by repression of enzyme synthesis.

The interesting and important regulatory properties of the *E. coli* aspartate carbamyltransferase have focused much attention on this enzyme; however, it is apparent that these properties are not typical of carbamyltransferases in all bacterial species. Three classes of aspartate carbamyltransferase have been recognized in other bacterial species; these differ in molecular size and kinetic characteristics (*29, 30*). The three known Class A enzymes appear to have molecular weights between 400,000 and 500,000, show sigmoidal kinetics, and are inhibited by nucleotides; these may be exemplified by the enzyme from *Pseudomonas fluorescens*. The Class B enzymes have molecular weights in the order of 300,000 and also display sigmoidal kinetics; the *E. coli* enzyme is assigned to this class. Class C carbamyltransferases are smaller, with molecular weights of about 100,000 to 140,000; the *S. faecalis* enzyme, which may be representative of this class, shows hyperbolic kinetics and is not subject to nucleotide regulation (*30*).

b. Aspartate Carbamyltransferase of Animal Cells. This activity has been demonstrated in a number of animal tissues and has been shown to be widely distributed in rat tissues (*31*); the enzyme has not been highly purified. This enzyme, along with carbamyl phosphate synthetase II (the glutamine-requiring enzyme which participates in pyrimidine synthesis, see below) and dihydroorotase, is found in the soluble fraction of cells; in contrast, ornithine carbamyltransferase and carbamyl phosphate synthetase I (both of which take part in urea synthesis) are mitochondrial enzymes and thus are in a different cell compartment.

Although aspartate carbamyltransferase appears to be a control point in the synthesis of the pyrimidine nucleotides in *E. coli*, the corresponding enzyme activity in animal cells has quite different properties and appears not to be a regulatory enzyme. In various animal tissues, the activity of this enzyme is in great excess (two to three orders of magnitude) over that of the initial step in uridylate synthesis, the formation of carbamyl phosphate by carbamyl phosphate synthetase II (*32, 33*). Thus, the transferase is not the rate-limiting step in this pathway and, accordingly, is not a likely regulatory site. Further to this point, the aspartate carbamyltransferase activity of various animal tissues [hemopoietic mouse spleen (*33*), rat liver (*34*), and frog eggs (*35*)] is unaffected by CTP and other pyrimidine nucleoside phosphates.

3. Carbamyl Phosphate Synthesis

In addition to the requirement for pyrimidine nucleotide synthesis, carbamyl phosphate is required for synthesis of arginine and urea. Carbamyl phosphate synthesis is a prominent activity in ureotelic liver and is aimed primarily at the formation of urea; the process of urea synthesis is served by a special carbamyl phosphate synthetase which is quite distinct from the enzymes responsible for carbamyl phosphate synthesis in extrahepatic tissues and in the livers of uricotelic animals. A third mechanism for synthesis of carbamyl phosphate is found in bacteria.

a. Carbamyl Phosphate Synthetase I. The carbamyl phosphate synthetase of a ureotelic liver (adult frog) was the first to be highly purified (*36*); the following reaction was catalyzed:

$$NH_4^+ + HCO_3^- + 2\ ATP \xrightarrow{\text{N-acetylglutamate}} H_2NCOOPO_3^{2-} + 2\ ADP + 2\ H^+ + P_i$$

This reaction is irreversible and requires two molecules of ATP; *N*-acetylglutamate is required in this reaction as an allosteric cofactor. Glutamine will *not* replace ammonia in this reaction. The enzyme will be referred to as carbamyl phosphate synthetase I.

The carbamyl phosphate synthetase I from mitochondria of rat liver has been partly purified and found to possess the same substrate requirements as the frog liver enzyme and to require *N*-acetylglutamate (*37*). Neither enzyme was able to utilize glutamine as a substrate.

It appears that, under certain circumstances, the carbamyl phosphate product from the liver enzyme may be diverted from its principal fate, urea synthesis, into the orotate pathway. This was evident in the work of Kesner (*38*), who demonstrated that the fasted rat excretes orotate in response to the administration of ammonium salts, but upon the administration of urea cycle intermediates, the flow of carbamyl phosphate is directed back into urea synthesis.

Carbamyl phosphate synthesis by the above mechanism (carbamyl phosphate synthetase I) is confined to rat liver mitochondria, a fact which at first seemed puzzling in view of very convincing evidence that the orotate pathway of pyrimidine biosynthesis functioned in many tissues from which the Type I synthetase was absent. It was evident that some other system for the generation of carbamyl phosphate must exist. The discovery of a glutamine-requiring carbamyl phosphate synthetase in mushrooms by Levenberg (39) suggested that glutamine, rather than ammonia, might be the primary nitrogen donor; similar enzymes were then found in E. coli, yeast, and in several animal tissues. This enzyme will be referred to as carbamyl phosphate synthetase II.

  b. *Carbamyl Phosphate Synthetase II.* The Type II synthetase from *E. coli* has been partly purified and the reaction established as follows (40):

$$\text{Glutamine} + \text{HCO}_3^- + 2\text{ATP} \xrightarrow[\text{Mg}^{2+}]{\text{K}^+} \text{glutamate} + \text{H}_2\text{NCOOPO}_3^{2-} + 2\text{ADP} + \text{P}_i$$
(or ammonia)

This enzyme differs from the Type I enzyme in that *both* glutamine and ammonia are substrates and *N*-acetylglutamate does not affect the rate. Anderson and Meister (40) have provided evidence that the first stage in this reaction is the ATP-dependent formation of an enzyme-bound form of carbon dioxide (see Chapter 5). The activity of the *E. coli* enzyme is subject to regulatory control by certain nucleotides; uridylate is inhibitory, but ornithine and inosinate activate the enzyme (41). The *E. coli* enzyme reversibly aggregates in the presence or absence of the positive effectors, ornithine or inosinate, but not in the presence of uridylate; however, manifestation of the positive allosteric effects does not depend upon the state of monomer–oligomer equilibrium. Oligomer formation is not required for catalytic activity (42).

Piérard et al. (43) have demonstrated that the synthesis of carbamyl phosphate synthetase II in *E. coli* is under repressor control and that a single species of this enzyme serves both the arginine and pyrimidine pathways. Uridylate is a negative effector of the synthetase and ornithine reverses this inhibition (44). The existence of two separate Type II synthetases has been demonstrated in yeast, one each serving the arginine and pyrimidine pathways (45).

Carbamyl phosphate synthetase II has also been found in the Ehrlich ascites carcinoma, pigeon liver, and in livers of young rats (32). Demonstration of this activity in the latter tissue was greatly facilitated by the fact that carbamyl phosphate synthetase I is absent prior to days 17–19 of gestation, whereupon its activity rises dramatically. Glutamine is likely to be the physiological substrate for this enzyme in animal cells; ammonia is toxic to animals, and plasma concentrations remain at low levels (for example, ammonium ion concentrations in human plasma ordinarily would not exceed $4 \times 10^{-5} M$).

The enzyme from hematopoietic mouse spleen has been purified about 90-fold by Tatibana and Ito (*33*); the following stoichiometry was demonstrated:

$$2\ ATP + glutamine + HCO_3^- + H_2O \longrightarrow 2\ ADP + carbamyl\ phosphate + P_i + glutamate$$

dATP or other triphosphates could not replace ATP as a substrate. Both glutamine and ammonia were substrates for the spleen enzyme, the activity of which was unaffected by *N*-acetylglutamate.

As in other reactions in which either ammonia or the amide group of glutamine is utilized (amination of UTP to form CTP, synthesis of NAD from desamido-NAD, and of guanylate from xanthylate), the carbamyl phosphate synthetase II reaction is inhibited by the glutamine analogues, azaserine and diazo-oxo-norleucine (see Chapter 5).

  c. Bacterial Carbamate Kinase.   In a third type of reaction, quite different from the two described above, carbamyl phosphate is formed in *S. faecalis* by the enzyme carbamate kinase (*46*):

$$H_2NCOO^- + ATP \longleftrightarrow H_2NCOOPO_3^{2-} + ADP$$

Carbamate, formed nonenzymatically from ammonia and carbon dioxide, is phosphorylated by ATP. The reaction is reversible and, when operating in the direction of carbamyl phosphate degradation, is exergonic (*46*). The utilization of carbamyl phosphate by certain bacterial species (obtained by degradation of citrulline, creatinine, and allantoin) as an energy source is possible by way of this enzyme (*47*).

III. Origin of the Cytidine Phosphates

  The cytidine phosphates do not have an independent origin *de novo;* they are derived from the uridine phosphates by an amination which occurs at the triphosphate level:

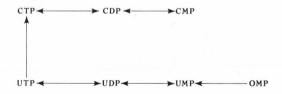

A. Cytidine Triphosphate Synthesis

Lieberman (48) demonstrated that a soluble enzyme preparation from E. coli converted UTP to CTP in a reaction which used ATP and ammonia, but not glutamine. Hurlbert and Kammen (49) were the first to demonstrate the synthesis of cytidine phosphates in preparations from animal cells; their enzyme preparations were treated to remove nucleotides and low molecular weight solutes and showed a requirement for glutamine, Mg^{2+}, ATP, and guanosine phosphates.

Subsequently, it was shown (50) that an extract from E. coli would utilize glutamine or ammonia as the amino donor (the difference from earlier work apparently being due to changes occurring during the preparation of the enzyme). Diazo-oxo-norleucine inhibited the glutamine-supported reaction, but did not block ammonia utilization. The partly purified bacterial enzyme had a specific requirement for nonstoichiometric amounts of GTP.

Long and Pardee (51) purified CTP synthetase from E. coli about 300-fold and established that when glutamine was a substrate, the presence of GTP was required as an allosteric effector. The activity of the enzyme was regulated in a complex manner by both substrates and by the product, CTP.

It has been concluded that CTP synthetase requires the following: Mg^{2+}, ATP, GTP, and glutamine or NH_4^+. The physiological amide donor in animal cells is undoubtedly glutamine. The reaction stoichiometry is

$$UTP + ATP + glutamine \xrightarrow[\text{GTP}]{\text{Mg}^{2+}} CTP + ADP + glutamate + P_i$$

B. Conversion of Cytidine Phosphates to Uridine Phosphates

The existence of deoxycytidylate deaminase is well established (Chapter 15), but cytidylate is not a substrate for this enzyme. There is no evidence for the enzymatic deamination of cytidine phosphates. The only means known for the direct conversion of cytidine compounds to uridine compounds is by way of deamination of cytidine itself; cytidine deaminase is discussed in Chapter 12.

IV. Regulation of the Orotate Pathway

A. General

Factors and conditions which influence the catalytic and kinetic characteristics of isolated enzymes are generally interpreted as regulatory mechanisms and are presumed to govern metabolic traffic in living cells—whether this is actually so is unproven.

Gerhart and Pardee (24) have pointed out that intracellular concentrations of ATP and CTP in *E. coli* are sufficiently high that these nucleotides could very well influence the activity of aspartate carbamyltransferase. However, this is not a regulatory enzyme in all bacterial species and it would appear unlikely to be a regulatory site in animal cells because the transferase of several animal cell types is insensitive to pyrimidine nucleotides and, furthermore, the transferase is clearly not a rate-limiting step in the orotate pathway of animal cells.

Bresnick (*52*) has concluded that in the partially hepatectomized rat, during the first 12 hours of liver regeneration, the activity of orotate phosphoribosyltransferase determines the rate of synthesis of the uridine phosphates. The potential activity of orotidylate decarboxylase is in excess of that of orotate phosphoribosyltransferase. This probably accounts for the virtual absence of orotidylate in animal tissues. The decarboxylation step is irreversible and may be subject to feedback inhibition by uridylate, which is a competitive inhibitor of the enzyme in rat liver and yeast (*19*).

A rare, recessive, human disease, hereditary orotic aciduria, is characterized by a megaloblastic anemia, retardation of growth, and the excretion of large quantities of orotic acid in the urine. The latter finding was explained by the discovery that individuals homozygous for the mutation have a deficiency in the activity of two enzymes of the orotate pathway, orotate phosphoribosyltransferase and orotidylate decarboxylase. These enzymatic deficiencies are manifested in erythrocytes and in cultured skin fibroblasts from homozygous patients (*53*) and evidently underlie the disease manifestations, because the latter are ameliorated by the administration of uridine. The genetic defect in this disease appears to involve *regulation* of enzyme synthesis because the activities of the two enzymes, which are absent from cells of homozygous individuals, return when these cells are cultured in the presence of the antimetabolite, azauridine (*54*) (see below).

B. AZAURIDINE

This compound has considerable growth inhibitory activity toward animal cells and, at one time, was regarded as having potential value for the chemotherapy of human cancer; however, clinical testing proved otherwise.

Azauridine is the 1-β-D-ribofuranoside of an unsymmetrical triazine dione:

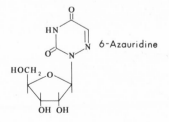

Azauridine is a potent inhibitor of the proliferation of some mouse tumors and of animal cells in culture, apparently acting as an inhibitor of the orotate pathway (55). Animals (including humans), when treated with this drug, excrete large amounts of orotate and orotidine, suggesting that the drug blocks orotidylate metabolism (56).

In cells, azauridine is converted to the 5'-monophosphate, which is a competitive inhibitor of orotidylate decarboxylase. Orotidylate accumulates because of the inhibited enzyme and its subsequent degradation explains the urinary excretion of orotidine in azauridine-treated animals. Azauridine has been used to treat acute leukemia in man, but resistance to the drug develops rapidly.

TABLE 11–I

ENZYMES OF PYRIMIDINE RIBONUCLEOTIDE BIOSYNTHESIS *de Novo*[a]

	Enzyme Commission number	Systematic name	Trivial name
	2.7.2.5	ATP: carbamate phospho-transferase (dephosphorylating)	Carbamyl phosphate synthetase I
1	—	—	Carbamyl phosphate synthetase II
	2.7.2.2	ATP: carbamate phosphotransferase	Carbamate kinase
2	2.1.3.2	Carbamoylphosphate: L-aspartate carbamoyl-transferase	Aspartate carbamyltransferase
3	3.5.2.3	L-4,5-Dihydro-orotate amidohydrolase	Dihydroorotase
4	1.3.3.1	L-4,5-Dihydro-orotate: oxygen oxidoreductase	Dihydroorotate dehydrogenase
5	2.4.2.10	Orotidine-5'-phosphate: pyrophosphate phosphoribosyltransferase	Orotate phosphoribosyltrans-ferase, orotidine-5'-phosphate pyrophosphorylase
6	4.1.1.23	Orotidine-5'-phosphate carboxy-lyase	Orotidine-5'-phosphate decarboxylase
7	6.3.4.2	UTP: ammonia ligase (ADP)	CTP synthetase

[a] Numbers in first column indicate reaction in summary diagram (p. 188).

V. Summary

The pathway of pyrimidine ribonucleotide biosynthesis *de novo* is summarized below. The enzymes involved are listed in Table 11–1.

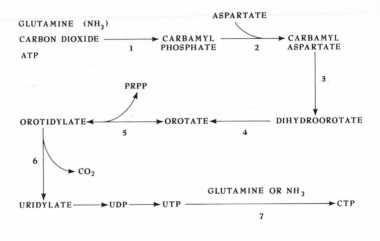

References

1. Hurlbert, R. B., and Potter, V. R., *J. Biol. Chem.* **195,** 257 (1952).
2. Hurlbert, R. B., and Potter, V. R., *J. Biol. Chem.* **209,** 1 (1954).
3. Reichard, P., *Advan. Enzymol.* **21,** 263 (1959).
4. Reichard, P., *in* "The Nucleic Acids" (E. Chargaff and J. N. Davidson, eds.), Vol. 2, p. 277. Academic Press, New York, 1955.
5. Kusama, K., and Roberts, E., *Biochemistry* **2,** 573 (1963).
6. Hager, S., and Jones, M. E., *J. Biol. Chem.* **240,** 4556 (1965).
7. Reichard, P., and Lagerkvist, U., *Acta Chem. Scand.* **7,** 1207 (1953).
8. Lieberman, I., and Kornberg, A., *J. Biol. Chem.* **207,** 911 (1954).
9. Bresnick, E., and Hitchings, G. H., *Cancer Res.* **21,** 105 (1961).
10. Smith, L. H., Jr., and Baker, F. A., *J. Clin. Invest.* **38,** 98 (1959).
11. Lieberman, I., and Kornberg, A., *Biochim. Biophys. Acta* **12,** 223 (1953).
12. Aleman, V., Handler, P., Palmer, G., and Beinert, H., *J. Biol. Chem.* **243,** 2560 (1968).
13. Hurlbert, R. B., and Reichard, P., *Acta Chem. Scand.* **8,** 701 (1954).
14. Lieberman, I., Kornberg, A., and Simms, E. S., *J. Biol. Chem.* **215,** 403 (1955).
15. Kasbekar, D. K., Nagabhushanam, A., and Greenberg, D. M., *J. Biol. Chem.* **239,** 4245 (1964).
16. Hatfield, D., and Wyngaarden, J. B., *J. Biol. Chem.* **239,** 2580 (1964).
17. Reyes, P., *Biochemistry* **8,** 2057 (1969).
18. Heidelberger, C., *Progr. Nucl. Acid Res. Mol. Biol.* **4,** 1 (1965).
19. Creasey, W. A., and Handschumacher, R. E., *J. Biol. Chem.* **236,** 2058 (1961).
20. Reichard, P., *Acta Chem. Scand.* **8,** 795 (1954).

21. Grisolia, S., and Cohen, P. P., *J. Biol. Chem.* **198**, 561 (1952).
22. Jones, M. E., Spector, F., and Lipmann, F., *J. Amer. Chem. Soc.* **77**, 819 (1955).
23. Cohen, P. P., and Marshal, M., *in* "The Enzymes" (P. Boyer, H. Lardy, and K. Myrbäck, eds.), 2nd rev. ed., Vol. 6, p. 327. Academic Press, New York, 1962.
24. Gerhart, J. C., and Pardee, A. B., *J. Biol. Chem.* **237**, 891 (1962).
25. Gerhart, J. C., and Schachman, H., *Biochemistry* **4**, 1054 (1965).
26. Weber, K., *Nature (London)* **218**, 1116 (1968).
27. Hammes, G. G., Porter, R. W., and Wu, C. W., *Biochemistry* **9**, 2992 (1970).
28. Yates, R. A., and Pardee, A. B., *J. Biol. Chem.* **227**, 677 (1957).
29. Bethell, M. R., and Jones, M. E., *Arch. Biochem. Biophys.* **134**, 352 (1969).
30. Prescott, L. M., and Jones, M. E., *Biochemistry* **19**, 3783 (1970).
31. Lowenstein, J. M., and Cohen, P. P., *J. Biol. Chem.* **220**, 57 (1956).
32. Hager, S. E., and Jones, M. E., *J. Biol. Chem.* **242**, 5674 (1967).
33. Tatibana, M., and Ito, K., *J. Biol. Chem.* **244**, 5403 (1969).
34. Bresnick, E., and Mossé, H., *Biochem. J.* **101**, 63 (1966).
35. Lan, S. J., Sallach, H. J., and Cohen, P. P., *Biochemistry* **8**, 3673 (1969).
36. Cohen, P. P., *Harvey Lect.* **60**, 119 (1966).
37. Kerson, L. A., and Appel, S. H., *J. Biol. Chem.* **243**, 4279 (1968).
38. Kesner, L., *J. Biol. Chem.* **240**, 1722 (1965).
39. Levenberg, B., *J. Biol. Chem.* **237**, 2590 (1962).
40. Anderson, P. M., and Meister, A., *Biochemistry* **4**, 2803 (1965).
41. Anderson, P. M., and Marvin, S. V., *Biochem. Biophys. Res. Commun.* **32**, 928 (1968).
42. Anderson, P. M., and Marvin, S. V., *Biochemistry* **9**, 171 (1970).
43. Piérard, A., Glandsdorff, N., Mergeay, M., and Wiame, J. M., *J. Mol. Biol.* **14**, 23 (1965).
44. Piérard, A., *Science* **154**, 1572 (1966).
45. Lacroute, F., Piérard, A., Grenson, M., and Wiame, J. M., *J. Gen. Microbiol.* **40**, 127 (1965).
46. Jones, M. E., and Lipmann, F., *Proc. Nat. Acad. Sci. U.S.* **46**, 1194 (1960).
47. Jones, M. E., *Science* **140**, 1373 (1963).
48. Lieberman, I., *J. Biol. Chem.* **222**, 765 (1956).
49. Hurlbert, R. B., and Kammen, H. O., *J. Biol. Chem.* **235**, 443 (1960).
50. Hurlbert, R. B., and Chakraborty, K. P., *Biochim. Biophys. Acta* **47**, 607 (1961).
51. Long, C. W., and Pardee, A. B., *J. Biol. Chem.* **242**, 4715 (1967).
52. Bresnick, E., *J. Biol. Chem.* **240**, 2550 (1965).
53. Howell, R. R., Klinenberg, J. R., and Krooth, R. S., *Johns Hopkins Med. J.* **120**, 81 (1967).
54. Pinsky, L., and Krooth, R. S., *Proc. Nat. Acad. Sci. U.S.* **57**, 925 (1967).
55. Handschumacher, R. E., and Welch, A. D., *in* "The Nucleic Acids" (E. Chargaff, and J. N. Davidson, eds.), Vol. 3, p. 453. Academic Press, New York, 1960.
56. Cardoso, S. S., Calabresi, P., and Handschumacher, R. E., *Cancer Res.* **21**, 1551 (1961).

CHAPTER 12

PYRIMIDINE RIBONUCLEOTIDE SYNTHESIS FROM BASES AND RIBONUCLEOSIDES

I. Introduction

It was apparent to early investigators that animal and microbial cells proliferated even without access to pyrimidines, and, therefore, were able to synthesize pyrimidines from simple precursors (Chapter 11). However, it was not evident whether preexisting pyrimidines or their derivatives could be utilized for nucleic acid synthesis. The discovery of bacterial mutants with absolute requirements for pyrimidines made it apparent that alternatives to the *de novo* synthetic route did exist, at least in microorganisms. Pyrimidine utilization was more difficult to demonstrate in animals. Because of the capability to synthesize pyrimidines *de novo*, the traditional type of nutrition experiment could not be employed and the matter remained unresolved until isotopically labeled pyrimidines became available. It first appeared that animals were not able to utilize free pyrimidines: Plentl and Shoenheimer in 1944 showed that ^{15}N-labeled uracil and thymine were *not* incorporated into rat liver nucleic acids, and in 1949, Bendich and Brown reported similar experiments showing that cytosine was not incorporated.

Although the free bases did not appear to be utilized, pyrimidine nucleosides were anabolized. In 1950, Hammarsten, Reichard and Saluste showed that labeled cytidine and uridine were incorporated into RNA and DNA of rat liver, and in 1951, Reichard and Estborn demonstrated that thymidine was incorporated into rat liver DNA (for a full account of early developments in this field, see references *1–4*).

It became apparent that of the free pyrimidines, uracil could be utilized by animals, earlier results notwithstanding. Lagerkvist *et al.* (*5*) showed that rapidly proliferating tissues of the rat, such as regenerating liver, intestinal mucosa, and hepatoma, would readily incorporate uracil into nucleic acids. Rapidly growing tissues have concentrations of pyrimidine-degrading enzymes that are low relative to those of the competing anabolic pathways. In adult rat liver, the enzymes of pyrimidine base catabolism are very active and degradation is the fate of low concentrations of uracil (see Chapter 13); however, uracil incorporation into nucleic acids does occur when uracil concentrations in liver are sufficiently high to saturate the catabolic enzymes (*6*).

Microorganisms utilize pyrimidine bases for incorporation into polynucleotides, and studies with bacteria and yeasts have been central in the elucidation of the enzymatic sequences of pyrimidine nucleotide metabolism (see review, reference *7*).

II. Metabolism of Pyrimidine Bases and Their Ribonucleosides

This discussion will be concerned mainly with uracil, cytosine, and their ribonucleosides; the metabolism of these compounds is summarized in Fig. 12-1. The metabolism of pyrimidine deoxyribonucleosides is considered in Chapter 14. Incorporation into the polynucleotides of uracil, cytosine, and their ribonucleosides proceeds by way of the ribonucleotide pool.

A. INTERCONVERSION OF URACIL AND CYTOSINE DERIVATIVES

Interconversions of the cytidine and uridine series are depicted in Fig. 12-1; it is apparent that only a few amination–deamination reactions link

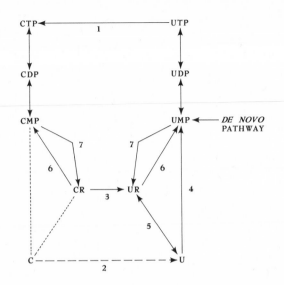

FIG. 12–1. Metabolism of uracil, cytosine, and their ribonucleosides. Solid lines, enzyme activity in both animal and microbial cells; large-dashed line, enzyme activity in microbial cells only; small-dashed lines, reaction not known. Reactions: (1) CTP synthetase; (2) cytosine deaminase; (3) cytidine deaminase; (4) uracil phosphoribosyltransferase; (5) uridine phosphorylase; (6) uridine-cytidine kinase; (7) 5′-nucleotidase.

these two series. The entire cytidine series is derived from the uridine series through a single reaction, the amination of UTP; this reaction, CTP synthetase (see Chapter 11), is the only known means by which uracil derivatives are converted into cytosine derivatives. Conversions in the other direction, i.e., deaminations, occur at the base or ribonucleoside levels, but not at the ribonucleotide level. That cytidine phosphates are not deaminated in *E. coli* and *S. typhimurium* is evident in the work of Neuhard *(8)*, who showed that mutants deficient in CTP synthetase, cytidine deaminase, and cytosine deaminase required both uracil and cytidine for growth.

1. Cytosine Deaminase (reaction 2, Fig. 12-1)

Cytosine deaminase is evidently absent from animal cells. Greenstein *et al.* *(9)* did not detect the deamination of cytosine in extracts of various rat and mouse tissues which, however, did actively deaminate cytidine.

Cytosine deaminase activity has been demonstrated in *E. coli* and yeast and has been partly purified from the latter source *(10)*; cytosine deaminase-deficient mutants have been isolated from various microorganisms *(7)*.

2. Cytidine Deaminase (reaction 3, Fig. 12-1)

This enzyme is found in various animal tissues and that from sheep liver has been purified 280-fold (11). Both cytidine and deoxycytidine are deaminated, as are the 5-halogen derivatives (e.g., 5-fluoro-deoxycytidine). Cytosine, cytidylate, and deoxycytidylate are not deaminated by this enzyme. Cytosine arabinoside (1-β-D-arabinofuranosylcytosine), which has an important use in the treatment of human neoplastic disease, is a substrate for cytidine deaminase; in the human, this compound is rapidly deaminated to yield the inactive uracil arabinoside (12). Cytidine deaminase is very active in E. coli and S. typhimurium, but is absent from yeasts and lactobacilli (7). The E. coli enzyme has been partly purified (13) and has a specificity similar to that of the animal enzyme. The E. coli enzyme deaminates 5-methyldeoxycytidine, enabling this compound to support the growth of a thymine-requiring E. coli mutant (14).

B. Uracil Phosphoribosyltransferase

Two routes are known by which the free base, uracil, can enter the ribonucleotide pool. One proceeds by the sequential actions of uridine phosphorylase and uridine-cytidine kinase (reactions 5 and 6, Fig. 12-1); this route is discussed below. The other route is by way of a single-step phosphoribosyltransferase reaction specific for uracil (reaction 4, Fig. 12-1):

$$\text{Uracil} + \text{PP-ribose-P} \rightarrow \text{uridylate} + \text{PP}_i$$

This enzyme has been demonstrated in several bacterial species and in yeast. The L. bifidus enzyme is distinct from orotate phosphoribosyltransferase from which it has been separated during partial purification. Cytosine is not a substrate for the E. coli enzyme, but 5-fluorouracil is accepted (15); the E. coli enzyme is activated by GTP and inhibited by UTP (16).

Uracil phosphoribosyltransferase has been demonstrated in the uracil-requiring protozoan, Tetrahymena pyriformis, and has been partly purified, but the substrate specificity of the preparation was not investigated (17).

The orotate phosphoribosyltransferase of yeast is specific for orotate and will not accept uracil as a substrate. This enzyme occurs in most animal cells as part of the de novo pathway of pyrimidine biosynthesis, but the specificity of the animal enzyme is unknown. Animal cells have a phosphoribosyltransferase activity capable of accepting pyrimidine substrates other than orotate, but it is not clear whether this is due to a phosphoribosyltransferase distinct from that of the orotate pathway (see Chapter 11).

Hatfield and Wyngaarden (18) have described a highly purified enzyme isolated from beef erythrocytes, which catalyzes the PP-ribose-P-dependent synthesis of nucleoside monophosphates from orotate, uracil, thymine, and the analogues, 5-fluorouracil and 6-azauracil; cytosine was not a substrate. This enzyme preparation also converted uric acid and xanthine into their 3-ribosyl 5'-monophosphate derivatives, evidently accepting the purines as 5,6-disubstituted uracil derivatives.

More recently, Reyes (*19*) has shown that cell-free extracts from cells of a transplantable rodent leukemia catalyze the PP-ribose-P-dependent synthesis of uridylate and 5-fluorouridylate; the enzyme involved, apparently a phosphoribosyltransferase, was virtually absent from a line of 5-fluorouracil-resistant mouse leukemia L1210 cells. Because of their ability to grow without this enzyme, the resistant cells must have obtained their pyrimidine nucleotide requirements by an alternate means, possibly by the phosphorylase–kinase route (reactions 5 and 6, Fig. 12-1), or by the *de novo* pathway.

In summary, a pyrimidine phosphoribosyltransferase activity is present in many cell types, animal and microbial; uracil and 5-fluorouracil, but not cytosine, are substrates for this enzyme.

C. Utilization of Pyrimidine Nucleosides

In cells, nucleoside molecules face two types of metabolic fate. In one, the *N*-glycosidic linkage will be cleaved; whether by phosphorolysis, hydrolysis, or glycosyl transfer depends upon the cell type and on the particular nucleoside involved. The alternate fate is phosphorylation to yield a nucleoside 5'-phosphate:

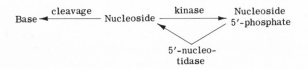

The kinase step enables cells to utilize nucleosides from their milieu for the incorporation into nucleotide coenzymes and polynucleotides, and has the obvious value that synthesis *de novo* is spared. The kinase reaction, operating in sequence with the readily reversible uridine phosphorylase reaction, also provides a route by which uracil may enter into the pyrimidine nucleotide pool.

1. *Uridine-Cytidine Kinase (reaction 6, Fig. 12-1)*

The many experiments in which labeled pyrimidine nucleosides have served as precursors of nucleoside phosphates and polynucleotides have inferred the existence of pyrimidine nucleoside kinases. Two pyrimidine deoxyribonucleoside kinases have been identified and isolated, thymidine kinase and deoxycytidine kinase (Chapter 14). Only one pyrimidine ribonucleoside kinase is known, that which phosphorylates uridine and cytidine according to the following reaction:

$$\text{Uridine} + \text{ATP} \xrightarrow{\text{Mg}^{2+}} \text{uridine 5'-phosphate} + \text{ADP}$$

Sköld (20) purified this enzyme 400-fold from Ehrlich ascites tumor cells and demonstrated the above stoichiometry and magnesium requirement. Various nucleosides were tested as substrates and the results were as follows:

Substrates	Nonsubstrates
UR, CR, 5-fluoro-UR,	TR, TdR, UdR
5-fluoro-CR, 6-azaUR	CdR, PuR, OR

It is noteworthy that both cytidine and uridine are substrates for this enzyme, whereas the corresponding deoxyribosides are not. There does not appear to be a separate cytidine kinase. It will be noted that orotidine is not a substrate for uridine kinase. Although orotidine is not ordinarily present in appreciable amounts in tissues, it appears, along with orotate, in the urine of animals treated with azauridine (see Chapter 11), and orotidine accumulates in the culture medium of certain *Neurospora* mutants which lack orotidylate decarboxylase. Orotidine is most likely a breakdown product of orotidylate and the provision in nature of an appropriate kinase activity would appear unnecessary.

Uridine-cytidine kinase from a rat hepatoma has been partly purified; this enzyme will accept several nucleoside triphosphates as substrates and appears to be subject to allosteric inhibition by CTP and UTP (21).

The uridine kinase of microorganisms has received little attention, although the activity has been demonstrated in several bacterial species and in yeasts (7).

2. Uridine Phosphorylase (reaction 5, Fig. 12-1)

Uridine and cytidine can be degraded in *Lactobacillus pentosus* (22) and *Saccharomyces cerevisiae* (23) by enzymes which effect hydrolytic cleavage of the base–sugar linkage. However, much more widely distributed in nature is the reversible phosphorolytic cleavage of this linkage. This is accomplished in animal cells and microorganisms by uridine phosphorylase, which catalyzes the following reaction:

$$\text{Uridine} + \text{P}_i \longleftrightarrow \text{Uracil} + \text{ribose-1-P}$$

Cytidine per se is not cleaved by this enzyme, nor by any known phosphorylase; however, cleavage does proceed following deamination by cytidine deaminase, which converts cytidine into uridine, the phosphorylase substrate.

The catabolism of uridine and cytidine by this route has value to the cell in that the ribose 1-phosphate produced may be degraded in energy-yielding reactions of the glycolytic pathway (see Chapter 6). Because of the ready reversibility of uridine phosphorolysis, this enzyme may contribute to the anabolism of uracil.

a. Uridine Phosphorolysis. The two pyrimidine nucleoside phosphorylases, uridine phosphorylase and thymidine phosphorylase, occur together in a number of cell types and have overlapping specificities with respect to cleavage of deoxyribonucleosides; these enzymes are readily separated by chromatography on DEAE cellulose (*24*). The specificity of uridine phosphorylase, isolated in this manner from various tissues, has been studied by Krenitsky *et al.* (*25*). Uridine phosphorylases in general have a considerable cleavage activity toward deoxyuridine and thymidine; uridine is the preferred substrate in a number of species (*25*), with cleavage activities toward deoxyuridine and thymidine diminishing in that order. This specificity was seen with a partly purified preparation of the enzyme from Ehrlich ascites carcinoma cells, which also cleaved 5-fluorouridine and 5-fluorodeoxyuridine; cytidine, deoxycytidine, and orotidine were not substrates (*26*).

The uridine phosphorylase of dog tissues is an exception to the above generalizations in that cleavage rates for deoxyuridine and thymidine are several times higher than for uridine. In spite of this preference for deoxyribosides, the enzyme is classified as a uridine phosphorylase because of several characteristics, including the elution position on chromatograms, pH dependence, and cleavage activity toward 3-ribosyl uric acid (*25*).

The uridine phosphorylase activities of animal tissues appears to be of two types, which can be distinguished on the basis of their pH optima. Uridine-cleaving enzymes of the chicken, human, guinea pig, and frog have pH optima in the vicinity of 6.5, as does the enzyme from *E. coli.* The tissues of the rat, mouse, and dog have uridine phosphorylase activities with pH optima about 8 (*25*).

Uridine phosphorylase, but not thymidine phosphorylase, is inhibited powerfully by the thymine nucleosides, deoxyxylosylthymine and deoxyglycosylthymine (*27*):

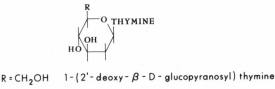

R = CH₂OH   1-(2'-deoxy- β - D - glucopyranosyl) thymine

R = H   1-(2'-deoxy- β - D - xylopyranosyl) thymine

b. Anabolic Function. Both purine and pyrimidine nucleoside phosphorylases catalyze freely reversible cleavage reactions. In the phosphorolysis of uridine, catalyzed by uridine phosphorylase from Ehrlich ascites carcinoma cells (*26*) and from *E. coli* (*28*), an equilibrium favoring nucleoside synthesis was reached at uracil:uridine ratios of about 1:4. The reversibilities of this reaction and of purine nucleoside phosphorylsis are the basis for reversible ribosyl transfer actions involving uridine phosphorylase which are discussed below. The reversibility of uridine phosphorolysis is central to the idea that this enzyme activity may participate in the anabolism of uracil.

Uridine phosphorylase participates in ribosyl transfer reactions which take place by the following mechanism. Ribose 1-phosphate produced by phosphorolysis of one nucleoside may serve as the ribose source in the phosphorylase-catalyzed synthesis of a different nucleoside:

$$B_1\text{-ribose} + \text{phosphate} \longleftrightarrow B_1 + \text{ribose 1-phosphate}$$

$$\text{Ribose 1-phosphate} + B_2 \longleftrightarrow B_2\text{-ribose} + \text{phosphate}$$

For example, intact Ehrlich ascites tumor cells, or extracts therefrom, transfer the ribosyl group of uridine to hypoxanthine and thereby catalyze the net synthesis of inosine; this reaction depends upon the coupled actions of uridine phosphorylase and purine nucleoside phosphorylase (*29*). Similar ribosyl transfers have been demonstrated with bacterial cells and extracts. Krenitsky has studied the kinetics of exchange between uracil-2-^{14}C and nonisotopic uridine catalyzed by highly purified uridine phosphorylase (*30*):

$$U^* + UR \longleftrightarrow U^*R + U$$

The appearance of ^{14}C in uridine was phosphate-dependent and maximum rates were not achieved until concentrations of uracil and phosphate were equivalent. An ordered mechanism was indicated involving the initial addition of pyrimidine or nucleoside; free ribose 1-phosphate is an intermediate in this mechanism.

The reversibility of uridine phosphorolysis suggests that uridine phosphorylase activity of the cell may be able to operate in the direction of synthesis, utilizing endogenous ribose 1-phosphate, and thereby contribute to the anabolism of uracil. Thus, uridine phosphorylase and uridine kinase acting in sequence would elevate uracil to the nucleotide level.

In summary, a catabolic function for uridine phosphorylase is clear. However, this enzyme may also participate in the anabolism of uracil, which is the evident function of uracil phosphoribosyltransferase. The operation of these apparently alternative routes of uracil anabolism in living cells has not yet been evaluated.

TABLE 12–I

ENZYMES OF PYRIMIDINE RIBONUCLEOTIDE SYNTHESIS
FROM BASES AND RIBONUCLEOSIDES[a]

Enzyme Commission number	Systematic name	Trivial name
1. 6.3.4.2	UTP: ammonia ligase (ADP)	CTP synthetase
2. 3.5.4.1	Cytosine aminohydrolase	Cytosine deaminase
3. 3.5.4.5	Cytidine aminohydrolase	Cytidine deaminase
4. —	—	Uracil phosphoribosyltransferase
5. 2.4.2.3	Uridine: orthophosphate ribosyltransferase	Uridine phosphorylase
6. 2.7.1.48	ATP: uridine 5′-phospho-transferase	Uridine-cytidine kinase
7. 3.1.3.5	5′-Ribonucleotide phospho-hydrolase	5′-Nucleotidase
— 2.4.2.1	Purine nucleoside: orthophos-phate ribosyltransferase	Purine nucleoside phosphorylase

[a] Numbers in first column refer to reactions in summary diagram (p. 192).

III. Summary

The pathways by which pyrimidine bases and ribonucleosides are metabolized are summarized in Fig. 12-1. The enzymes involved are listed in Table 12–I.

References

1. Schulman, M. P., in "Chemical Pathways of Metabolism" (D. M. Greenberg, ed.), 1st ed., Vol. 2, p. 223. Academic Press, New York, 1954.
2. Reichard, P., in "The Nucleic Acids" (E. Chargaff and J. N. Davidson, eds.), Vol. 2, p. 277. Academic Press, New York, 1955.
3. Reichard, P., Advan. Enzymol. 21, 263 (1959).
4. Crosbie, G. W., in "The Nucleic Acids" (E. Chargaff and J. N. Davidson, eds.), Vol. 3, p. 323. Academic Press, New York, 1960.
5. Lagerkvist, U., Reichard, P., Carlsson, B., and Grabosz, J., Cancer Res. 15, 164 (1955).
6. Canellakis, E. S., J. Biol. Chem. 227, 329 (1957).
7. O'Donovan, G. A., and Neuhard, J., Bacteriol. Rev. 34, 278 (1970).
8. Neuhard, J., J. Bacteriol. 96, 1519 (1968).

9. Greenstein, J. P., Carter, C. E., Chalkley, H. W., and Leuthhardt, F. M., *J. Nat. Cancer Inst.* **7**, 9 (1952).
10. Kream, J., and Chargaff, E., *J. Amer. Chem. Soc.* **74**, 4274 (1952).
11. Wisdom, G. B., and Orsi, B. A., *Eur. J. Biochem.* **7**, 223 (1969).
12. Camiener, G. W., and Smith, C. G., *Biochem. Pharmacol.* **14**, 1405 (1965).
13. Wang, T. P., Sable, H. Z., and Lampen, J. O., *J. Biol. Chem.* **184**, 17 (1950).
14. Cohen, S. S., and Barner, H. D., *J. Biol. Chem.* **226**, 631 (1957).
15. Brockman, R. W., Davis, J. M., and Stutts, P., *Biochim. Biophys. Acta* **40**, 22 (1960).
16. Malloy, A., and Finch, L. R., *FEBS Lett.* **5**, 211 (1969).
17. Heinrikson, R. L., and Goldwasser, E., *J. Biol. Chem.* **239**, 1177 (1964).
18. Hatfield, D., and Wyngaarden, J. B., *J. Biol. Chem.* **239**, 2580 (1964).
19. Reyes, P., *Biochemistry* **8**, 2057 (1969).
20. Sköld, O., *J. Biol. Chem.* **235**, 3273 (1960).
21. Orengo, A., *J. Biol. Chem.* **244**, 2204 (1969).
22. Lampen, J. O., and Wang, T. P., *J. Biol. Chem.* **198**, 385 (1952).
23. Grenson, M., *Eur. J. Biochem.* **11**, 249 (1969).
24. Krenitsky, T. A., Barclay, M., and Jacquez, J. A., *J. Biol. Chem.* **239**, 805 (1964).
25. Krenitsky, T. A., Mellors, J. W., and Barclay, R. K., *J. Biol. Chem.* **240**, 1281 (1965).
26. Pontis, H., Degerstedt, G., and Reichard, P., *Biochim. Biophys. Acta* **51**, 138 (1961).
27. Etzold, G., Preussel, B., and Langen, P., *Mol. Pharmacol.* **4**, 20 (1968).
28. Paege, L. M., and Schlenk, F., *Arch. Biochem. Biophys.* **40**, 42 (1952).
29. Paterson, A. R. P., and Simpson, A. I., *Can. J. Biochem.* **43**, 1693 (1965).
30. Krenitsky, T. A., *J. Biol. Chem.* **243**, 2871 (1968).

CATABOLISM OF PYRIMIDINE BASES

I. Introduction

The catabolism of pyrimidine nucleotides, like that of purine nucleotides (Chapter 10), involves dephosphorylation, deamination, and glycosidic bond cleavage. In contrast to purine catabolism, however, the pyrimidine bases are most commonly subjected to reduction rather than to oxidation. An oxidative pathway is found in some bacteria however.

So far as is known, pyrimidine ribo- and deoxyribonucleotides are dephosphorylated by the nucleotidases and phosphatases described in Chapter 10 as acting on purine nucleotides. Although a number of potential dephosphorylating enzymes may be available in most cells, the relative quantitative importance of each is not known.

The deamination of cytosine and its derivatives, and the cleavage of the glycosidic bond of uridine, deoxyuridine, and thymidine, are discussed in detail elsewhere (Chapters 11, 12, 14, and 15). Here, therefore, only the breakdown of the pyrimidine bases uracil and thymine will be discussed, together with a brief survey of urinary end products of pyrimidine metabolism. The reactions of pyrimidine base catabolism were worked out some years ago, and primary references may be found in the reviews of Potter (*1*) and Schulman (*2*).

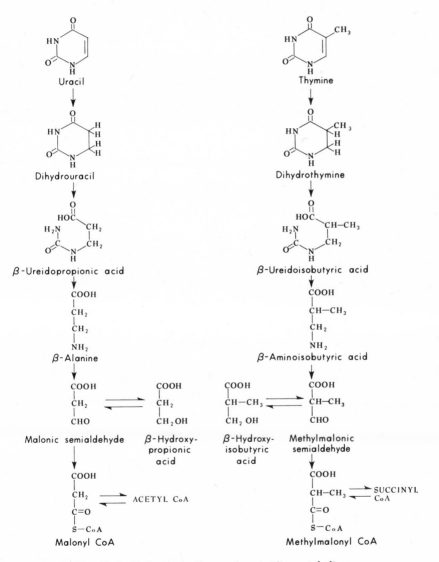

FIG. 13-1. Reductive pathway of pyrimidine catabolism.

II. The Reductive Pathway

Figure 13-1 shows that reduction of the 5:6 double bond is the first step in the breakdown of the pyrimidine ring. This is followed by oxidative cleavage of the 3:4 bond to form ureidopropionic or ureidoisobutyric acid, from which β-amino acids are formed with the release of NH_3 and CO_2. Whether carbamate or carbamyl phosphate is the primary product is not known. The β-amino acids are then oxidatively deaminated to β-aldehydes, which are converted to acids which are variously metabolized.

This is the pathway of pyrimidine base breakdown followed in animals, and it is apparent that these compounds may be catabolized completely to carbon dioxide and ammonia. Small amounts of β-alanine and β-amino-isobutyrate are excreted in the urine, but they may also be formed from certain amino acids.

5-Halouracils are metabolized by similar reactions. 5-Bromo- and 5-iodouracil are dehalogenated at the reductive step, whereas 5-fluorouracil is not. The latter analogue is converted to the α-fluoro analogue of β-ureido-propionic acid, but this compound is not further metabolized. 6-Azauracil is not a substrate for these reactions and is excreted unchanged.

Rat liver catabolizes uracil at a high rate, and, as has been mentioned in Chapter 12, this base is consequently not readily incorporated into nucleic acids. However, as shown by the data in Table 13-I (3), by increasing the concentration of uracil the catabolic enzyme system of rat liver slices can be saturated, and significant amounts are then incorporated into RNA.

TABLE 13-I

EFFECT OF CONCENTRATION ON THE
RELATIVE AMOUNT OF URACIL
ANABOLIZED AND CATABOLIZED
IN RAT LIVER SLICES[a]

Uracil (μM)	Percent conversion to	
	RNA	CO_2
0.518	0.0045	98
2.0	0.034	90
8.0	0.11	45
30.0	0.35	22

[a] From (3).

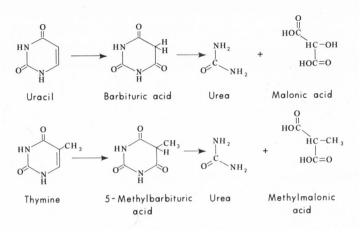

FIG. 13-2. Oxidative pathway of pyrimidine catabolism.

III. Urinary Pyrimidines

Weissmann *et al.* (*4*) have calculated that in man approximately 800 to 1000 mg of uracil are synthesized *de novo* per day. The normal urinary excretion of orotate is approximately 1.4 mg per day, and that of orotidine, 2.5 mg per day (*5*). In patients with orotic aciduria due to decreased orotidylate decarboxylase, the excretion of orotate may increase 20-fold (*5*), although the excretion of orotidine does not necessarily increase.

Trace amounts of uracil, uridine, thymine, and cytosine may also be found in human urine, along with pseudouridine, 5-methylcytosine, 3-methylcytosine, and 2'-O-methylcytidine (*6, 7*). Immediately following X-irradiation the urinary excretion of deoxycytidine and thymidine is markedly increased (*8*).

TABLE 13-II

CATABOLIC ENZYMES OF PYRIMIDINE METABOLISM

Enzyme Commission number	Systematic name	Trivial name
1.3.1.1	4,5-Dihydro-uracil:NAD oxido-reductase	Dihydro-uracil dehydrogenase
1.3.1.2	4,5-Dihydro-uracil:NADP oxido-reductase	Dihydro-uracil dehydrogenase (NADP)
1.2.99.1	Uracil:(acceptor oxidoreductase)	Uracil dehydrogenase

IV. The Oxidative Pathway

Certain bacteria can oxidize uracil to barbituric acid (4-oxouracil) and thymine to the corresponding 5-methylbarbituric acid (see Fig. 13-2). An enzyme called barbiturase cleaves the pyrimidine ring to form malonic acid and urea, probably through a ureidomalonic acid intermediate, although this has not been identified.

The enzymes of pyrimidine catabolism are listed in Table 13–II.

References

1. Potter, V. R., "Nucleic Acid Outlines," p. 217. Burgess, Minneapolis, Minnesota, 1960.
2. Schulman, M. P., *Metab. Pathways* **2**, 389 (1961).
3. Canellakis, E. S., *J. Biol. Chem.* **227**, 701 (1957).
4. Weissmann, S. M., Eisen, A. Z., Fallon, H., Lewis, M., and Karon, M., *J. Clin. Invest.* **41**, 1546 (1962).
5. Latz, M., Fallon, H. J., and Smith, L. H., Jr., *Nature (London)* **197**, 194 (1963).
6. Heirweigh, K. P. M., Ramboer, C., and DeGroote, J., *Amer. J. Med.* **42**, 913 (1967).
7. Fink, K., and Adams, W. S., *Arch. Biochem. Biophys.* **126**, 27 (1968).
8. Guri, C. D., and Cole, L. J., *Clin. Chem.* **14**, 383 (1968).

PART IV

PURINE AND PYRIMIDINE
DEOXYRIBONUCLEOTIDE METABOLISM

The deoxyribonucleotides are quantitatively minor constituents of the acid-soluble fraction of tissues, being present in concentrations roughly two orders of magnitude lower than the ribonucleotides. A gram of fresh rat liver contains about 0.05 μmole of free deoxyribonucleoside derivatives; in contrast, the ribonucleotide content is about 8 μmoles/g. Over half of this deoxyribonucleosidic material is in the form of pyrimidine derivatives; of these, deoxycytidine compounds are most abundant.

Deoxyribonucleic acid synthesis is the principal fate of the deoxyribonucleotides, and cells in a nonproliferating state apparently do not contain appreciable pools of these compounds. DNA is made during a particular portion of the intermitotic period (the S phase of the cell cycle) and it appears that the free deoxyribonucleotide precursors of DNA are made mainly at this time and are incorporated into the polynucleotide with little accumulation. This, coupled with the fact that only a small proportion of cells in most animal tissues are undergoing division at any one time, accounts for the low concentration of deoxyribonucleotides in animal tissues.

The deoxyribonucleotides are derived primarily from the ribonucleotides. There is no evidence for a deoxyribosidic homologue of PP-ribose-P and, hence, phosphodeoxyribosyltransferase reactions analogous to those in ribonucleotide metabolism do not contribute to deoxyribonucleotide formation. Synthesis of deoxyribonucleotides by reduction of ribonucleotides takes place at the diphosphate or triphosphate level, depending on the organism (Chapter 16).

Deoxyribonucleosides may be available from dietary constituents or may arise from the breakdown of endogenous deoxyribonucleotidic material. Enzymatic capabilities exist for the catabolism of deoxyribonucleosides, and as well, for their conversion to deoxyribonucleotides (Chapter 14).

At the nucleotide level, the metabolism of purine deoxyribonucleotides is uncomplicated; adenine and guanine deoxyribonucleotides are not interconverted, and hydrolysis or transfer reactions involving phosphoryl groups are the only major transformations known.

Metabolism of the pyrimidine deoxyribonucleotides is more complex because, in addition to transfer of phosphoryl groups, deamination and methylation reactions occur at this level. Specifically, the thymidine phosphates are derived by methylation of deoxyuridylate, and the latter may be derived from the deoxycytidine phosphates by way of deoxycytidylate deaminase. The deoxycytidine phosphates are not formed by amination of deoxyuridine phosphates, but are derived entirely from the cytidine phosphates by enzymatic reduction (Chapter 16).

The following diagram illustrates in very broad terms routes by which cellular deoxyribonucleoside phosphates are derived and utilized, and is intended to provide perspective for the more detailed discussion of Chapters 14–16.

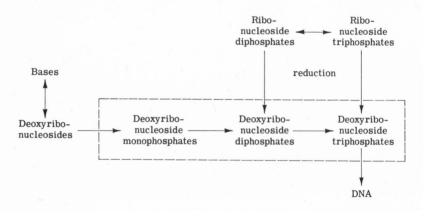

CHAPTER 14

DEOXYRIBONUCLEOSIDE METABOLISM

I. Introduction

Two metabolic fates are open to deoxyribonucleosides: elevation to the
nucleotide level by phosphorylation or cleavage with release of the free

207

base and utilization of the deoxyribosyl moiety. The phosphorolytic cleavage of a deoxyribonucleoside yields deoxyribose 1-phosphate, which may be degraded by energy-yielding pathways or utilized to form other deoxyribonucleosides.

The reversibility of the phosphorylases would appear to afford a route for incorporation of free bases into the deoxyribonucleotide pool:

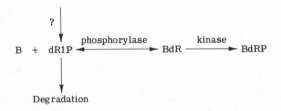

However, for adenine, guanine, and uracil, the dominant route of anabolism is by way of their *ribo*nucleotide derivatives and traffic along the deoxyribosidic route is not ordinarily significant. Because cytosine is not a substrate for nucleoside phosphorylases, incorporation by the phosphorylase–kinase route is not possible for this base. The other pyrimidine base of DNA, thymine, is poorly anabolized by both animal and bacterial cells, in spite of the fact that most cells possess thymidine phosphorylase, the action of which is readily reversible. This suggests that ordinarily cellular supplies of deoxyribose 1-phosphate are not available for base anabolism. Experiments are cited below in which it was demonstrated that a significant contribution to the biogenesis of deoxyribose of DNA in *E. coli* cells did not occur by a route other than ribonucleotide reduction.

Although cells derive their deoxyribonucleotides primarily by reduction of ribonucleotides, the deoxyribonucleosides can be utilized to some extent by way of kinase reactions. Phosphorylation of deoxyribonucleosides represents the only known point of entry into the sequences of deoxyribonucleotide metabolism other than the main entry point, ribonucleotide reduction. Pyrimidine deoxyribonucleosides may be incorporated into DNA in animal cells by this route. The conversion of deoxyadenosine and deoxyguanosine into deoxyribonucleotides and thence into DNA would also appear possible by a kinase-initiated sequence, because the enzymatic phosphorylation of these compounds has been demonstrated. However, this route has not been well studied in animal cells and its assessment is complicated by very active deamination and phosphorolytic cleavage reactions which compete with the reactions leading to DNA. *E. coli* cells appear to possess only one kinase capable of phosphorylating deoxyribonucleosides, thymidine kinase.

II. Deoxyribonucleoside Deamination

Cytidine deaminase is responsible for the deamination of both cytidine and deoxycytidine:

$$CR \rightarrow UR + NH_3$$

$$CdR \rightarrow UdR + NH_3$$

The enzyme from *E. coli* (*1*) also deaminates 5-methyldeoxycytidine (see Chapter 12). Deoxyadenosine is a substrate for adenosine deaminase (see Chapter 10), but deoxyguanosine is not known to be deaminated.

III. Deoxyribonucleoside Phosphorolysis

The phosphorolytic cleavage of deoxyribonucleosides is reversible and analogous to that of the ribonucleosides.

$$B + dR1P \longleftrightarrow BdR + P_i$$

Deoxyribonucleoside phosphorolysis is catalyzed by purine nucleoside phosphorylase, thymidine phosphorylase, and to some extent by uridine phosphorylase; the specificity of animal uridine phosphorylases differs with the cells of origin (see Chapter 12).

A. PHOSPHOROLYSIS OF PURINE DEOXYRIBONUCLEOSIDES

The purine nucleoside phosphorylase activity of animal tissues cleaves both ribo- and deoxyribonucleosides of 6-oxypurines. For example, the the purine nucleoside phosphorylase of human erythrocytes has been highly purified and crystallized by Parks and co-workers (*2*); this enzyme will cleave the ribo- and deoxyribonucleosides of guanine and hypoxanthine. Zimmerman *et al.* (*3*) have shown that purine nucleoside phosphorylase of several animal tissues has a low intrinsic activity toward adenine in the presence of ribose 1-phosphate. The cleavage of deoxyadenosine by highly purified preparations of the animal enzyme has not been reported; but by analogy with adenosine, one might expect it also to be a poor substrate. The specificities of the purine nucleoside phosphorylases of *E. coli* and *S. typhimurium* differ from that of the animal enzyme in that adenosine and deoxyadenosine are readily phosphorolyzed (*4–6*).

As mentioned above, deoxyadenosine is a substrate for adenosine deaminase, which can convert it from a poorly phosphorolyzed compound into the rapidly cleaved deoxyinosine. By these means, cells are equipped for the catabolism of purine deoxyribonucleosides.

The following diagram summarizes purine deoxyribonucleoside phosphorolysis:

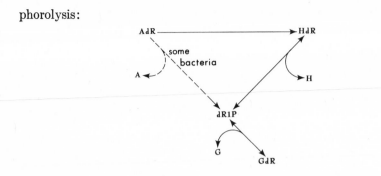

B. PHOSPHOROLYSIS OF PYRIMIDINE DEOXYRIBONUCLEOSIDES

Deoxyuridine and thymidine are substrates for pyrimidine nucleoside phosphorylases, but deoxycytidine (and cytidine) is generally regarded as being inert to phosphorolysis (7); Tarr's demonstration of deoxycytidine formation from cytosine in extracts of fish milt is an exception to this generalization (8). Catabolism of cytosine nucleosides is initiated by deamination to form uracil nucleosides which can be phosphorolyzed.

Although uridine is the preferred substrate of the uridine phosphorylase of animal tissues, deoxyuridine and thymidine are also cleaved at appreciable rates. The uridine phosphorylase activity of dog tissues readily cleaves deoxyuridine and thymidine and, in fact, may be the only means by which the pyrimidine deoxyribonucleosides are phosphorolyzed in dog tissues, because thymidine phosphorylase is absent from a number of them (9) (see Chapter 12). Uridine phosphorylase of *E. coli* is highly specific toward the ribosyl portion of its substrate and cleavage of deoxyribonucleosides is slower relative to uridine than with the animal enzymes (9).

Thymidine phosphorylase cleaves deoxyuridine and certain derivatives with substituents in the 5-position, including thymidine. This enzyme is of irregular distribution, notably being absent from some animal tissues and tumors (9). The enzyme has been purified from *E. coli* and from several mouse tissues and is distinct from uridine phosphorylase (9–11); both the bacterial and animal enzymes cleave deoxyuridine. This enzyme should probably be referred to as thymidine-deoxyuridine phosphorylase.

A summary follows:

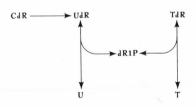

C. Deoxyribonucleoside Catabolism by Phosphorolysis

Phosphorolytic cleavage of both ribo- and deoxyribonucleosides is a process of value to cells in that sugar phosphates are released from the nucleosidic linkage and these can then be utilized in energy-yielding degradative sequences. Deoxyribose 1-phosphate and ribose 1-phosphate, generated by nucleoside phosphorolysis, may be converted to the corresponding 5-phosphates by the action of phosphoribomutase (4). Ribose 5-phosphate can be metabolized by way of the pentose intermediates of the glycolytic pathway (see Chapter 6), and deoxyribose 5-phosphate can be degraded by deoxyriboaldolase to yield acetaldehyde and the glycolytic intermediate, glyceraldehyde 3-phosphate (see below). In contrast to the ready metabolism of pentose phosphates, free pentoses are not well utilized in most animal cells. The pathways of deoxyribonucleoside catabolism are shown in Fig. 14-1.

D. Anabolic Function of Deoxyribonucleoside Phosphorolytic Activity

A means of converting free bases into deoxyribonucleotides would appear to be afforded by the sequential action of nucleoside phosphorylases and kinases; however, cells have only a very limited capability for the endogenous formation of the deoxyribose 1-phosphate needed for this

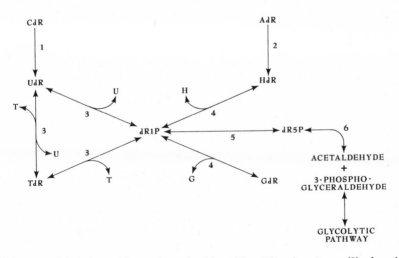

Fig. 14-1.   Catabolism of deoxyribonucleosides: (1) cytidine deaminase; (2) adenosine deaminase; (3) thymidine phosphorylase; (4) purine nucleoside phosphorylase; (5) phosphoribomutase; (6) deoxyriboaldolase.

sequence. In the presence of exogenous deoxyribonucleosides, through the activities of nucleoside phosphorylases, deoxyribosyl groups attached to one base may be transferred to another base; such reactions are of real significance in bacteria, particularly in lactobacilli.

1. Reversal of Phosphorolysis

The reversibility of deoxyribonucleoside phosphorolysis suggests that bases might be elevated to the deoxyribonucleotide level by successive actions of a phosphorylase and a kinase:

$$B + dR1P \underset{\text{phosphorylase}}{\xrightleftharpoons{\hspace{2cm}}} BdR \xrightarrow{\text{kinase}} BdRP$$

The first step, deoxyribonucleoside synthesis, has been demonstrated *in vitro* with purified preparations of purine ribonucleoside phosphorylase and thymidine phosphorylase (*11–13*). However, it is uncertain whether a significant incorporation of base occurs in cells by this route, without an extracellular source of deoxyribosyl groups (see below). The possibility that deoxyribose 1-phosphate may be generated endogenously by way of deoxyriboaldolase is discussed in Section IV.

Experiments with animal cells have shown that the pyrimidine bases are much less effective DNA precursors than the corresponding deoxyribonucleosides, although the interpretation of such experiments is complicated by the rapid catabolism of uracil and thymine which takes place in liver. The incorporation of thymine into DNA in animals (*14*), or in *in vitro* systems (*15*) is slow and contrasts with the much more rapid incorporation of thymidine.

2. Deoxyribosyl Transfer via Free Deoxyribose 1-Phosphate ("Coupled" Deoxyribosyl Transfer)

Purine nucleoside phosphorylase and thymidine phosphorylase have been shown to catalyze transfer of the deoxyribosyl group from one base to another by reaction sequences that involve the intermediate formation of free deoxyribose 1-phosphate by the following mechanism:

$$B_1dR + P_i \rightleftharpoons B_1 + dR1P \tag{1}$$

$$B_2 + dR1P \rightleftharpoons B_2dRP + P_i \tag{2}$$

$$\text{Net:} \quad B_1dR + B_2 \rightleftharpoons B_1 + B_2dR \tag{3}$$

Deoxyribose 1-phosphate, derived from initial phosphorolysis of deoxyribonucleoside B_1dR, is utilized to form the new nucleoside B_2dR, the net result (3) being deoxyribosyl transfer. This sequence is phosphate-dependent.

In purine–purine deoxyribosyl transfer, a single enzyme, purine nucleoside phosphorylase (see Chapter 10), participates in both reactions (1) and (2). For example, the purified enzyme catalyzes the following reaction (*16*):

$$G + HdR \longleftrightarrow GdR + H \tag{4}$$

By catalyzing both reactions (1) and (2) above, thymidine phosphorylase will transfer the deoxyribosyl group between thymine and uracil (note that cytosine is not a substrate). For example, the following transfer has been demonstrated with thymidine phosphorylase of human leukocytes (*17*):

$$T + UdR \longleftrightarrow TdR + U \tag{5}$$

As noted above, uridine phosphorylases of some animal tissues have an appreciable activity toward deoxyribonucleosides (*9*) and, accordingly, may participate in transfer reactions of this sort.

Purine–pyrimidine deoxyribosyl transfer reactions result when reactions (1) and (2) are catalyzed by the joint actions of purine nucleoside phosphorylase and thymidine phosphorylase, that is, when the activities of these enzymes are coupled. For example, the following reaction is catalyzed by extracts of human leukocytes (*18*):

$$T + GdR \longleftrightarrow TdR + G \tag{6}$$

3. "Direct" Transfer of Deoxyribosyl Groups by Thymidine Phosphorylase

In addition to participation in the deoxyribosyl transfer reactions described above, in which *free* deoxyribose 1-phosphate is formed as an intermediate, thymidine phosphorylase also catalyzes deoxyribosyl transfers involving thymine and uracil in which deoxyribosyl phosphate is an intermediate but is enzyme-bound (*18–20*). Such transfers require nonstoichiometric amounts of phosphate (*19*). The reaction mechanisms of uridine phosphorylase and purine nucleoside phosphorylase are not of this type and, accordingly, "direct" deoxyribosyl transfers occur only between substrates for thymidine phosphorylase, as exemplified by reaction (5) above and the following (*19*) (the asterisk indicates ^{14}C-labeling):

$$T^* + TdR \longleftrightarrow T^*dR + T$$

Thymidine phosphorylases of both animal cells and *E. coli* catalyze this "direct" transfer of deoxyribosyl groups between thymine and uracil.

4. Metabolism of Thymine and Thymidine in E. coli

Coupling of purine nucleoside phosphorylase and thymidine phosphorylase in proliferating *E. coli* cells is demonstrated in the data of Fig.

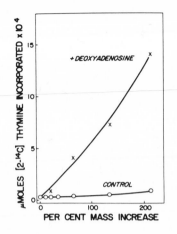

FIG. 14-2. Incorporation of exogenous thymine-2-^{14}C into acid-insoluble material by exponentially growing cells of *Escherichia coli* (wild type) in presence and absence of deoxyadenosine (0.2 mM) (data of A. Munch-Petersen, cited in reference *4*). Reproduced with permission.

14-2; this experiment measured the effect of exogenous deoxyadenosine on the incorporation of thymine-2-^{14}C into acid-insoluble material (a measure of incorporation into DNA). It is apparent in Fig. 14-2 that free thymine is incorporated at a very low rate in *E. coli*, as is the case for most bacteria. However, if deoxyadenosine is provided in the medium, an extensive incorporation occurs; deoxyinosine and deoxyguanosine also stimulate thymine incorporation into DNA (*21*). It appears that thymine anabolism is ordinarily limited by the endogenous provision of deoxyribose 1-phosphate, but the latter can be derived by phosphorolysis of exogenous deoxyribonucleosides. As noted above, purine–pyrimidine deoxyribosyl transfer is not direct and requires the intermediate formation of free deoxyribose 1-phosphate.

Thymidine phosphorylase occurs in most microorganisms; however, its absence from lactobacilli is noteworthy (*4*). The *E. coli* enzyme has been partly purified (*9–11*) and will phosphorolyze deoxyuridine derivatives with various substituents at the pyrimidine 5-position, but will not accept those with a 4-amino group. Thymidine phosphorylase is quantitatively released from *E. coli* cells by osmotic shock, a result which is thought to mean that the enzyme is located near the cell surface (*21*). When wild-type *E. coli* cells are provided with thymidine, initially the deoxyribonucleoside is utilized for DNA synthesis. However, this soon ceases because the synthesis of thymidine phosphorylase is induced and, as a result, thymidine is cleaved to the free base and the sugar phosphate degraded.

5. Deoxyribonucleoside Transfer in Lactobacilli

Many *Lactobacillus* species have an absolute growth requirement for a purine, a pyrimidine, and a single deoxyribonucleoside of any sort; because the entire complement of cellular deoxyribonucleotides is derived from these materials, there is an evident requirement for the capability of transferring the deoxyribosyl function between bases in these cells. However, in lactobacilli deoxyribosyl transfer is not accomplished by the phosphorylase mechanisms outlined above; in this connection it may be noted that lactobacilli are devoid of thymidine phosphorylase (4). A specific enzyme activity, *trans*-N-deoxyribosylase, catalyzes deoxyribosyl transfer in these bacteria; this is accomplished without the intermediate formation of deoxyribose 1-phosphate, is readily reversible, and takes place in the absence of inorganic phosphate:

$$B_1dR + B_2 \longleftrightarrow B_1 + B_2dR$$

This enzyme activity, which is clearly different from that of the nucleoside phosphorylases, was discovered by MacNutt (22) in *Lactobacillus helveticus*. Beck and Levin (23) have found the enzyme in other lactobacilli that depend on a single exogenous deoxynucleoside for growth; synthesis of the enzyme was under repressor control in these strains. The enzyme can be prepared free of nucleoside phosphorylase activity (24).

This enzyme has not been highly purified and there is some evidence that there may be two or more enzymes involved, with one being specific for purine–purine transfers. Specificity for deoxyribonucleosides has been shown clearly. It is noteworthy that cytosine and deoxycytidine are substrates for the *trans*-N-deoxyribosylase; this is convincing evidence of an identity separate from the phosphorylases because deoxycytidine is not a substrate for the latter.

IV. Sources of Deoxyribose

As already stated, cells would appear to have the means of converting bases into deoxyribonucleotides through the reversibility of nucleoside phosphorolysis, providing deoxyribose 1-phosphate was available. It will be seen below that the coupled activities of deoxyriboaldolase and phosphoribomutase (enzymes involved in degradation of deoxyribose phosphates) will generate deoxyribose 1-phosphate *in vitro* under certain conditions and, therefore, these enzymes represent a possible cellular source of this compound. However, it has not been demonstrated that any contribution to the formation of deoxyribonucleoside phosphates is made by this route *in vivo* and, furthermore, evidence cited below indicates that *E. coli* cells are able to derive deoxyribonucleotides by a means entirely independent of the endogenous generation of deoxyribose 1-phosphate.

Thus, while a contribution to deoxyribonucleotide synthesis by way of deoxyriboaldolase may be postulated, none has been shown and ribonucleotide reduction appears to be the primary means by which deoxyribosyl groups are formed.

A. DEOXYRIBOALDOLASE

In 1952, Racker described an enzymatic activity in *E. coli* and in animal tissues which catalyzed the reversible condensation of glyceraldehyde 3-phosphate and acetaldehyde to yield deoxyribose 5-phosphate.

$$\text{Acetaldehyde} + \text{D-glyceraldehyde 3-phosphate} \longrightarrow \text{2-deoxy-D-ribose 5-phosphate}$$

This reaction was recognized as a potential intracellular source of deoxyribose 1-phosphate because the mutase-catalyzed interconversion of 1- and 5-phosphate esters of deoxyribose was known. Boxer and Shonk (*25*) showed that extracts of rat liver formed deoxyribose phosphate when incubated with hexose diphosphate and threonine; it was evident that the threonine aldolase activity of the extracts provided acetaldehyde for the *in vitro* synthesis of deoxyribose phosphate and it was postulated that this might also occur *in vivo:*

$$\text{Threonine} \longrightarrow \text{glycine} + \text{acetaldehyde}$$

The deoxyriboaldolase of rat liver has been highly purified and shown to have an absolute requirement for a carboxylate ion activator, such as citrate (*26*). Groth and Jiang (*27*) found that *meso-α,β*-diphenyl succinate inhibited this enzyme, acting competitively with citrate. In Ehrlich ascites cells, diphenyl succinate inhibited the incorporation of labeled adenine into DNA purines, but not into RNA purines. This suggests participation of the aldolase in the conversion of adenine into DNA through provision of the deoxyribose moiety of dAMP. However, it has not been demonstrated directly that a pathway involving deoxyriboaldolase contributes to deoxyribonucleotide synthesis and further, evidence cited below argues against this route as a deoxyribosyl source, but does not preclude the possibility of its operation.

While an anabolic role for deoxyriboaldolase is speculative, its catabolic function has been demonstrated (*28*); this enzyme affords a means of utilizing carbohydrate from deoxyribonucleosides for energy production.

B. FORMATION OF PYRIMIDINE DEOXYRIBONUCLEOTIDES EXCLUSIVELY
    FROM RIBONUCLEOTIDES

That uridine or cytidine can be the exclusive precursor of the pyrimidine deoxyribonucleotide subunits of DNA was shown by Karlström and Larsson (29) with a pyrimidine-requiring *E. coli* mutant, OK305. After growing the bacteria in the presence of uridine or cytidine totally labeled with ^{14}C, DNA and RNA were isolated and degraded to their constituent nucleotides. The specific activities of the sugar and base portions of the nucleotides from RNA and DNA were compared (Tables 14–I and 14–II).

In the data of Table 14–I it is seen that when uridine was the sole pyrimidine source, the specific activities of the sugar portions of the RNA pyrimidine nucleotides were the same as those derived from DNA. This means that the pyrimidine deoxyribonucleotides were synthesized exclusively from the same precursors as the pyrimidine ribonucleotides. There was some reduction in the isotope content of the pentose, but none in that of the base; this apparently came about through exchange with cellular nonisotopic ribose phosphates by way of the reversible reaction of uridine phosphorylase.

TABLE 14–I

SPECIFIC ACTIVITIES OF NUCLEOTIDES ISOLATED FROM DNA AND RNA OF
Escherichia coli OK305 AFTER GROWTH IN
MEDIA CONTAINING URIDINE-^{14}C[a]

	Specific activities (counts/min/μmole) of			
Nucleotide (or nucleoside)	Base	Pentose	Pentose : base ratio
Uridine added to the medium	20,600	9,300	11,300	1.22[b]
AMP	710	—	—	—
CMP	11,700	9,400	2,300	0.24
GMP	720	—	—	—
UMP	11,500	9,700	2,800	0.32
dAMP	200	—	—	—
dCMP	11,900	9,200	2,700	0.29
dGMP	250	—	—	—
dTMP	11,200	9,200	2,000	0.22

[a] From (29). Reproduced with permission.
[b] Theoretical value: 1.25.

TABLE 14-II

SPECIFIC ACTIVITIES OF NUCLEOTIDES ISOLATED FROM DNA AND RNA OF
Escherichia coli OK305 AFTER GROWTH IN A MEDIUM
CONTAINING CYTIDINE-[14]C[a]

	Specific activities (counts/min/μmole) of			
	Nucleotide (or nucleoside)	Base	Pentose	Pentose : base ratio
Cytidine added to the medium	17,300	7,500	9,800	1.31[b]
AMP	430	—	—	—
CMP	16,900	7,600	9,300	1.22
GMP	340	—	—	—
UMP	8,200	7,700	500	0.06
dAMP	200	—	—	—
dCMP	17,100	7,400	9,700	1.31
dGMP	200	—	—	—
dTMP	14,700	7,500	7,200	0.96

[a] From (*29*). Reproduced with permission.
[b] Theoretical value: 1.25.

When cytidine was the sole pyrimidine source (Table 14–II), poly-nucleotide cytidylate and deoxycytidylate had identical specific activities and were incorporated without dilution. This is convincing evidence that in *E. coli*, reduction of ribonucleoside derivatives is able to supply all of the deoxyribose for DNA components without appreciable contribution from other sources, such as the postulated sequence involving deoxyriboaldolase mentioned above. It will be recognized that this result does not preclude the possibility that alternate routes of deoxyribosyl biosynthesis may operate in circumstances other than the special ones of the Karlström and Larsson experiment. The details of ribonucleotide reduction are discussed at length in Chapter 16.

V. Deoxyribonucleoside Kinases

A. UTILIZATION OF DEOXYRIBONUCLEOSIDES

In proliferating cells, deoxyribonucleosides may be utilized for DNA synthesis following an initial phosphorylation by nucleoside triphosphate-dependent kinases:

$$\text{BdR} \xrightarrow{\text{kinase}} \text{BdRP} \rightarrow \rightarrow \text{BdRPPP} \rightarrow \text{DNA}$$

Mammalian cells (30) and lactobacilli (31) have this general capability.

In animal cells thymidine is specifically incorporated into DNA thymidylate, contributing neither to DNA deoxycytidylate nor to the pyrimidine ribonucleotides (32). Since incorporation into DNA is the only major anabolic fate of labeled thymidine, DNA synthesis can be measured without isolation of DNA—only the removal of unused substrate and acid-soluble metabolites is necessary. Tritiated thymidine has been particularly valuable in autoradiographic studies of DNA synthesis and is available today in very high specific activities.

In animal cells engaged in DNA synthesis, deoxyuridine can be incorporated without cleavage into DNA thymidylate, but phosphorolysis also occurs and the uracil so released will not have incorporation into DNA as a specific metabolic fate. Because of deamination, deoxycytidine may be converted to either of the pyrimidine nucleotides of DNA; cleavage of the deamination product, deoxyuridine, liberates uracil which may enter the pathways of pyrimidine ribonucleotide metabolism.

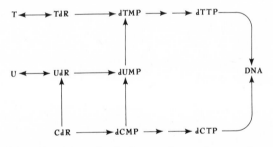

The deoxyribonucleosides of adenine and guanine are incorporated into the DNA purines in animal cells to some degree, but extensive cleavage of the N-glycosidic bond occurs with subsequent reutilization of the base by the ribonucleotide pathways.

In contrast to the general capabilities of mammalian cells and lactobacilli for utilizing deoxyribonucleosides, E. coli cells can utilize thymidine for DNA synthesis, but are unable to directly utilize deoxyadenosine, deoxyguanosine, and deoxycytidine, because the appropriate kinases are absent (33).

B. PURINE DEOXYRIBONUCLEOSIDE KINASES

Deoxyadenosine is phosphorylated in animal cells; for example, Klenow (34) has shown that Ehrlich ascites tumor cells incubated with deoxyadenosine accumulate substantial quantities of deoxyadenosine phosphates, of which dATP is the most abundant. The kinase responsible for the initial phosphorylation has not been identified, but may well be deoxycytidine kinase, an enzyme of broad specificity (see below). Both deoxyadenosine and deoxyguanosine are substrates for the deoxycytidine kinase of calf thymus; in a 500-fold purified preparation of this enzyme (35, 36) a single molecular species was responsible for the phosphorylation of deoxycytidine, deoxyguanosine, and deoxyadenosine. The protein catalyzing these phosphorylations showed coincident purification, the substrates were mutually inhibitory, and dCTP was a powerful inhibitor of the phosphorylation of the three nucleosides. Studies of the allosteric regulation of the activity of this enzyme toward the purine deoxyribonucleosides have not been reported. For deoxyadenosine, the apparent Michaelis constant of the calf thymus enzyme is 7×10^{-5} M, about 10 times that for deoxycytidine.

Adenosine kinase has a low activity toward deoxyadenosine; the apparent Michaelis constant of the partly purified enzyme from rabbit liver for deoxyadenosine (2×10^{-3} M) is about 1000 times that for adenosine (37).

Whether there are other kinase activities, perhaps specific for the purine deoxyribonucleosides, remains an open question. The phosphorylation of deoxyguanosine has been demonstrated [for example, in fish milt extracts (38)], but the enzymes responsible have not been identified, apart from the ability of deoxycytidine kinase to catalyze this reaction. The apparent Michaelis constant for the phosphorylation of deoxyguanosine by the calf liver kinase is 3.1×10^{-4} M, about two orders of magnitude higher than that for deoxycytidine (36).

C. DEOXYCTIDINE KINASE

1. General

The kinase which converts deoxycytidine to its 5′-monophosphate has been studied most extensively in preparations from calf thymus (35, 36). The preferred substrate is deoxycytidine, for which the Michaelis constant (5×10^{-6} M) is much lower than that of two other substrates, deoxyadenosine and deoxyguanosine. Cytidine, uridine, and thymidine are not phosphorylated by this enzyme. Deoxycytidine kinase is subject to a complex pattern of allosteric regulation by nucleotides. The end product of deoxycytidine phosphorylation, dCTP, is a potent inhibitor; this inhibition is reversed by dTTP. The enzyme has a rather broad specificity for the phosphate donor, with the triphosphates of the natural ribo- and deoxyribonucleosides being substrates; the inactivity of dCTP is a notable exception.

Deoxycytidine kinase activity is distinct from uridine-cytidine kinase, which is a separate enzyme and does not accept deoxyribonucleosides as substrates.

Lymphoid tissues of the rat and the mouse have high levels of deoxycytidine kinase activity relative to other tissues (39), suggesting that the utilization of exogenous deoxycytidine may be important in these tissues. In this connection it may be noted that animals have appreciable concentrations of deoxycytidine in the circulation; the concentration of deoxycytidine in rat blood is about 0.04 mM (40). The presence of deoxycytidine in the blood may be related to the fact that cytosine nucleosides are not subject to phosphorolytic cleavage, as noted above.

2. Cytosine Arabinoside

Cytosine arabinoside, an analog of deoxycytidine, is a substrate for deoxycytidine kinase and is a competitive inhibitor of the phosphorylation of the natural substrate; the Michaelis constant for the calf thymus enzyme is 10 times higher than that of deoxycytidine (35). Cytosine arabinoside inhibits the growth of a variety of experimental neoplasms and has an important use in the chemotherapy of human cancer. The lethal effects of the analog toward malignant and other cells are generally attributed to inhibition of DNA synthesis. Cytosine arabinoside is extensively anabolized in cells; the mono-, di-, and triphosphate derivatives are formed and some incorporation into internal nucleotides in DNA occurs. The drug-induced inhibition of DNA synthesis has been attributed to the competitive inhibition of DNA polymerase by cytosine arabinoside triphosphate (41, 42). A necessary first step in this mechanism of action would be the kinase-catalyzed phosphorylation of cytosine arabinoside; the limited activity of deoxycytidine kinase in resistant mouse lymphoma cells is consistent with this mechanism (43). The ability of deoxycytidine to protect cells against the inhibitory effects of cytosine arabinoside is presumably due in some measure to the competition between deoxycytidine and the arabinoside for the catalytic site of deoxycytidine kinase. Cytosine arabinoside is deaminated by cytidine deaminase to form the pharmacologically inactive uracil arabinoside.

D. THYMIDINE KINASE

This enzyme has received a great deal of attention because its activity in cells and tissues is related to their proliferative state. Thymidine kinase activity is high in cells engaged in DNA synthesis; for example, it is elevated in regenerating liver.* Twenty-four hours after partial hepatectomy in rats, the thymidine kinase content of the liver remnant is elevated six-fold, whereas the activities of deoxyadenosine and deoxycytidine kinases are much less affected (*44*). In the isoproterenol-stimulated salivary gland, thymidine kinase activity is also increased. About 20 hours after the injection of isoproterenol into rats and mice, a wave of mitotic activity occurs in the salivary glands and a large increase in the rate of DNA synthesis occurs simultaneously (*45*). Thymidine kinase activity increases markedly and then subsides, in coincidence with these events. In cultured animal cells, thymidine kinase activity varies with the stage of the mitotic cycle. The enzyme activity is low immediately after division, rises during the period of DNA synthesis, and falls rapidly during the subsequent division (*46*).

An additional feature of thymidine kinase activity is that it is subject to allosteric regulation. Despite the implication in these facts that this enzyme activity is important in the deoxyribonucleotide economy of cells, thymidine is not an obligatory intermediate in the biosynthesis of the thymidine phosphates and the mutational loss of this enzyme is not lethal.

The *E. coli* enzyme has been purified 1200-fold by Okazaki and Kornberg (*47, 48*) who demonstrated the stoichiometry of the reaction:

$$\text{TdR} + \text{ATP} \rightarrow \text{dTMP} + \text{ADP}$$

In addition to thymidine, the enzyme will accept as substrates deoxyuridine and various 5-position derivatives including those with fluoro, bromo, and iodo substituents. The kinase is specific for deoxyribonucleosides and will not phosphorylate deoxycytidine.

The *E. coli* enzyme is subject to allosteric regulation; it is activated by by dCDP and dCTP, and inhibited by dTTP, but not by dTDP or thymidylate. In the presence of dCDP, the apparent Michaelis constant for thymidine decreases and the V_m increases; the sigmoidal nature of the rate—

---

* Following partial hepatectomy, or cellular damage, the surviving liver tissue proliferates, or "regenerates" to restore the loss. About 70% of the liver of the rat may be removed; cells of the remaining lobes divide until their mass approximates that of the original tissue, a process that takes about 3 weeks.

A series of changes takes place in the nucleic acid metabolism of the remnant, some of which are evident within an hour after surgery. A well-defined wave of DNA synthesis sweeps through the tissue 18–24 hours after surgery; this event coincides roughly with a burst of mitotic activity. Some synchronization of cell metabolic activities is indicated by the time course of these events; however, the synchrony is soon lost. Definition of the biochemical factors which initiate the sudden burst of DNA synthesis has been a particular goal of investigators.

ATP saturation curve changes to hyperbolic in the presence of activator. Both inhibitory and activating nucleotides appear to induce dimerization of the enzyme, a step thought to be necessary for the enzyme to assume active or inactive conformations in response to the nucleotide effectors (*49*).

Thymidine kinase has been purified about 100-fold from several rat tumors by Bresnick and Thompson (*50*), who showed a specific inhibition by dTTP.

The relative importance of thymidine kinase as part of the DNA-synthesizing apparatus is put into perspective by reports of mutant cell lines which lack this enzyme. Morris and Fischer (*51*) have described the deletion of thymidine kinase in strains of bacteria and cultured tumor cells which were resistant to 5-fluorodeoxyuridine. These cells were able to proliferate without a functional thymidine kinase.

We may conclude that this enzyme provides a means of utilizing thymidine (which is not an intermediate in thymidylate synthesis *de novo*, but which may occur endogenously or in animal diets); its synthesis and activity are subject to controls, apparently because formation of the product, thymidylate, is a key step in the regulation of DNA synthesis.

TABLE 14–III

Enzymes of Deoxyribonucleoside Metabolism[a]

Enzyme Commission number	Systematic name	Trivial name
1 2.4.2.4	Thymidine:orthophosphate deoxyribosyltransferase	Thymidine phosphorylase
2.4.2.6	Nucleoside:purine (pyrimidine) deoxyribosyltransferase	Nucleoside deoxyribosyl-transferase
2 2.7.1.2	ATP:thymidine 5'-phospho-transferase	Thymidine kinase
3 3.5.4.5	Cytidine aminohydrolase	Cytidine deaminase
4 —	—	Deoxycytidine kinase
5 —	—	Phosphoribomutase
6 4.1.2.4	2-Deoxy-D-ribose-5-phosphate acetaldehyde-lyase	Deoxyriboaldolase
7 2.4.2.1	Purine-nucleoside:orthophosphate ribosyltransferase	Purine nucleoside phosphorylase
8 —	—	*trans*-N-Deoxyribosylase
9 3.5.4.4	Adenosine aminohydrolase	Adenosine deaminase

[a] Numbers in first column refer to reactions in summary diagram (p. 224).

VI. Summary

The pathways by which purine and pyrimidine deoxyribonucleosides are metabolized are summarized below. The enzymes are listed in Table 14–III.

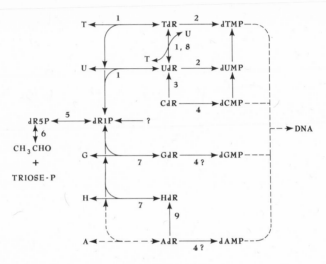

References

1. Wang, T. P., Sable, H. Z., and Lampen, J. O., *J. Biol. Chem.* **184**, 17 (1950).
2. Sheen, M. R., Kim, B. K., and Parks, R. E., Jr., *Mol. Pharmacol.* **4**, 293 (1968).
3. Zimmerman, T. P., Gersten, N. B., Ross, A. F., and Miech, R. P., *Can. J. Biochem.* **49**, 1050 (1971).
4. O'Donovan, G. A., and Neuhard, J., *Bacteriol. Rev.* **34**, 278 (1970).
5. Tazuke, Y., and Yamada, H., *Agr. Biol. Chem.* **27**, 625 (1963).
6. Robertson, B. C., Jargiello, P., Blank, J., and Hoffee, P. A., *J. Bacteriol.* **102**, 628 (1970).
7. DeVerdier, C. H., and Potter, V. R., *J. Nat. Cancer Inst.* **24**, 9 (1960).
8. Tarr, H. L. A., *Can. J. Biochem.* **42**, 51 (1964).
9. Krenitsky, T. A., Mellors, J. W., and Barclay, R. K., *J. Biol. Chem.* **240**, 1281 (1965).
10. Razzell, W. E., and Casshyap, P., *J. Biol. Chem.* **239**, 1789 (1964).
11. Razzell, W. E., and Khorana, H. G., *Biochim. Biophys. Acta* **28**, 562 (1958).
12. Tarr, H. L. A., and Roy, J. E., *Can. J. Biochem.* **45**, 409 (1967).
13. Gallo, R. C., Perry, S., and Breitman, T. R., *J. Biol. Chem.* **242**, 5059 (1967).
14. Brown, G. B., and Roll, P. M., *in* "The Nucleic Acids" (E. Chargaff and J. N. Davidson, eds.), Vol. 2, p. 341. Academic Press, New York, 1955.
15. Friedkin, M., and Wood, H., *J. Biol. Chem.* **220**, 639 (1956).
16. Krenitsky, T. A., *Mol. Pharmacol.* **3**, 526 (1967).
17. Gallo, R. C., and Perry, S., *J. Clin. Invest.* **48**, 105 (1969).

18. Gallo, R. C., and Breitman, T. R., *J. Biol. Chem.* **243**, 4936 (1968).
19. Krenitsky, T. A., *J. Biol. Chem.* **243**, 2871 (1968).
20. Zimmerman, M., and Seidenberg, J., *J. Biol. Chem.* **239**, 2622 (1964).
21. Kammen, H. O., *Biochim. Biophys. Acta* **134**, 301 (1967).
22. MacNutt, W. S., *Biochem. J.* **50**, 384 (1952).
23. Beck, W. S., and Levin, M., *J. Biol. Chem.* **238**, 702 (1963).
24. Roush, A. H., and Betz, R. F., *J. Biol. Chem.* **233**, 261 (1958).
25. Boxer, G. E., and Shonk, C. E., *J. Biol. Chem.* **233**, 535 (1958).
26. Groth, D. P., *J. Biol. Chem.* **242**, 155 (1967).
27. Groth, D. P., and Jiang, N., *Biochem. Biophys. Res. Commun.* **22**, 62 (1966).
28. Munch-Petersen, A., *Eur. J. Biochem.* **15**, 191 (1970).
29. Karlström, O., and Larsson, A., *Eur. J. Biochem.* **3**, 164 (1967).
30. Reichard, P., *Acta Chem. Scand.* **11**, 11 (1957).
31. Hoff-Jörgensen, E., *Biochem. J.* **50**, 400 (1951).
32. Reichard, P., and Estborn, B., *J. Biol. Chem.* **188**, 839 (1951).
33. Karlström, O., *Eur. J. Biochem.* **17**, 68 (1970).
34. Klenow, H., *Biochim. Biophys. Acta* **61**, 885 (1962).
35. Durham, J. P., and Ives, D. H., *J. Biol. Chem.* **245**, 2276 (1970).
36. Ives, D. H., and Durham, J. P., *J. Biol. Chem.* **245**, 2285 (1970).
37. Lindberg, B., Klenow, H., and Hansen, K., *J. Biol. Chem.* **242**, 350 (1967).
38. Tarr, H. L. A., *Can. J. Biochem.* **42**, 1535 (1964).
39. Durham, J. P., and Ives, D. H., *Mol. Pharmacol.* **5**, 358 (1969).
40. Rotherham, J., and Schneider, W. C., *J. Biol. Chem.* **232**, 853 (1958).
41. Graham, F. L., and Whitmore, G. F., *Cancer Res.* **30**, 2627 (1970).
42. Graham, F. L., and Whitmore, G. F., *Cancer Res.* **30**, 2636 (1970).
43. Chu, M. Y., and Fischer, G. A., *Biochem. Pharmacol.* **14**, 333 (1965).
44. Beltz, R. E., *Arch. Biochem. Biophys.* **99**, 304 (1962).
45. Whitlock, J. P., Jr., Kaufman, R., and Baserga, R., *Cancer Res.* **28**, 2211 (1968).
46. Brent, T. P., Butler, J. A. V., and Crathorn, A. R., *Nature (London)* **207**, 176 (1965).
47. Okazaki, R., and Kornberg, A., *J. Biol. Chem.* **239**, 269 (1964).
48. Okazaki, R., and Kornberg, A., *J. Biol. Chem.* **239**, 275 (1964).
49. Iwatsuki, N., and Okazaki, R., *J. Mol. Biol.* **29**, 155 (1967).
50. Bresnick, E., and Thompson, U. B., *J. Biol. Chem.* **240**, 3967 (1965).
51. Morris, N. R., and Fischer, G. A., *Biochim. Biophys. Acta* **68**, 84 (1963).

CHAPTER 15

INTERCONVERSION OF DEOXYRIBONUCLEOTIDES

I. Thymidylate Synthesis

A. INTRODUCTION

The formation of thymidine phosphates is a strategically important part of the complex of metabolic events which provide nucleotide building blocks for DNA replication. Thymidylate is a unique component of DNA and all four deoxyribonucleoside triphosphates must be available for the synthesis of DNA chains to proceed. This strategic importance and the complex array of regulatory controls acting on the enzymatic steps leading to thymidylate have led to suggestions that DNA synthesis may be influenced or even regulated through such controls. However, thymidine triphosphate has been implicated as an allosteric effector in the regulation of various steps in the metabolism of other nucleotides (for example, in the conversion of ribonucleotides to deoxyribonucleotides) and, accordingly, it may be more realistic at this time to view the regulatory controls on thymidylate metabolism as part of a complex set of control mechanisms that, through their interactions in the living cell, achieve a balanced production of nucleotides for polynucleotide synthesis and other purposes.

In the preceding chapter it was shown that thymidylate may be derived from exogenously supplied thymidine; however, most cells are independent of an external source of thymidine and are able to synthesize their own thymidylate from the endogenous metabolite, deoxyuridylate:

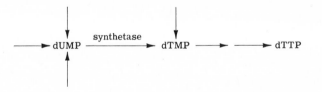

The several reactions leading to deoxyuridylate will be discussed subsequently (Section II).

B. THYMIDYLATE SYNTHETASE

Prior to the definitive experiments of Friedkin and Kornberg (see below), early tracer studies with orotate and with nucleoside derivatives of uracil and cytosine had demonstrated that the pyrimidine ring of DNA thymine could be derived from these compounds. In addition, it had become apparent that the methyl group of thymine was derived from "one-carbon units" at the oxidation level of formaldehyde rather than at the oxidation level of formate; formaldehyde and the hydroxymethyl group of serine were recognized as precursors of the thymine methyl group (*1*). Also, nutritional and biochemical experiments had shown that folic acid derivatives took part in the incorporation of the one-carbon units into the thymine methyl group (*2*). Finally, Friedkin and Kornberg (*3*) showed that cell-free extracts from *E. coli* would convert deoxyuridylate to dTTP in the presence of serine, ATP, and H_4-folate, and further, that the place of serine in this experiment could be taken by substrate amounts of synthetic "hydroxymethyl H_4-folate" (5,10-methylene H_4-folate; see Chapter 5).

It was apparent from the substrates and products of this reaction that the incoming one-carbon unit must undergo reduction to reach the oxidation level of the methyl group, and it was demonstrated that H_4-folate itself was the reductant, emerging from the reaction as H_2-folate (*4*). Although formaldehyde will serve as precursor of the thymidylate methyl group in enzymatic experiments, the hydroxymethyl group of serine is the probable physiological source, entering by way of the serine hydroxymethyltransferase reaction:

$$\text{Serine} + H_4\text{-folate} \rightarrow 5,10\text{-methylene } H_4\text{-folate} + \text{glycine}$$

It was evident that the by-product of thymidylate synthesis, H_2-folate, could be reduced to H_4-folate through activity of tetrahydrofolate dehydrogenase:

$$H_2\text{-Folate} + \text{NADPH} + H^+ \rightarrow H_4\text{-folate} + \text{NADP}^+$$

The sequential action of these three enzymes to form a "thymidylate-synthesizing cycle" was visualized, in which H_4-folate was regenerated (by tetrahydrofolate dehydrogenase), recharged with a one-carbon unit (by serine hydroxymethyltransferase), and returned to the reaction.

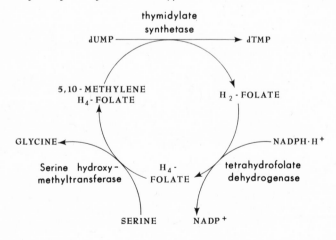

Thus, the formation of thymidylate involves the formation of a carbon–carbon bond and a reduction; 5,10-methylene H_4-folate is both donor of the one-carbon unit and reductant. It is apparent that a catalytic amount of H_4-folate will serve in the cycle. This is consistent with the observation that in *S. faecalis* extracts the synthesis of thymidylate in the presence of excess H_4-folate was unaffected by aminopterin, a potent inhibitor of tetrahydrofolate dehydrogenase; when this reaction was conducted with catalytic amounts of H_4-folate and excess NADPH, thymidylate synthesis was blocked by aminopterin (5).

The thymidylate synthetases from *E. coli* and *S. faecalis* have been extensively purified; as well, the reaction has been demonstrated in crude extracts from various animal tissues. This enzyme activity is low or not detectable in normal rat liver, but appears in regenerating liver in approximate coincidence with the surge of DNA synthesis which occurs 18–24 hours after partial hepatectomy. In general, the activity of thymidylate synthetase is low in cells not engaged in DNA synthesis.

It is well known that phage infection induces in the bacterial host cells the synthesis of new, phage-specified enzymes that are involved in the synthesis of DNA for the phage progeny. Thymidylate synthetase is one of a number of such enzymes which appear in *E. coli* cells upon infection with the T-even phages (see Section I, D). This was first recognized by Barner and Cohen (6), who showed that phage T2 proliferated in *E. coli* B_T^- (and, therefore, thymine-containing DNA was synthesized), even though cells of this mutant strain lack thymidylate synthetase activity and require thymine for growth. After infection with phage T2, thymidylate synthetase activity in the mutant cells increased 1000-fold; this new enzyme activity differed in a number of respects from the normal *E. coli* enzyme and so was attributed to the phage genome.

Thymidylate synthetase from chick embryo has been purified 600–800-fold (7). This enzyme was found to methylate uridylate at 40% of the rate for deoxyuridylate; the Michaelis constant for the latter was 10^{-2} to 10^{-3} times that for uridylate. No physiological significance is apparent for the methylation of uridylate. The chick embryo enzyme also showed slight activity toward deoxycytidylate, as did the enzyme from *S. faecalis*.

Regulatory control of bacterial thymidylate synthetases by nucleotides has not been reported; however, Lorenson *et al.* (7) investigated that of the chick embryo thymidylate synthetase and found that the 5'-mono-, di-, and triphosphates of adenosine, cytidine, guanosine, and uridine, and of their deoxyribosides, did *not* inhibit activity. Thus, the activity of the synthetase, from this source at least, does not seem to be subject to allosteric regulation by nucleotides. With crude extracts of Ehrlich ascites carcinoma cells, Reyes and Heidelberger (8) found that thymidylate was a weak product inhibitor; however, regulation by other nucleotides was not tested.

Studies on the mechanism of the bacterial thymidylate synthetase reaction employing tritiated H_4-folate showed that a hydrogen atom of 5,10-methylene H_4-folate was transferred to the thymine methyl group and suggested the formation of a methylene bridge between folate and deoxyuridylate (9).

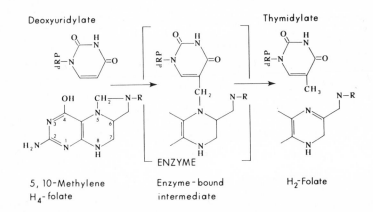

The adduct, which presumably is enzyme-bound, is thought to undergo an intramolecular rearrangement to form thymidylate and dihydrofolate, which then leave the enzyme. Experiments with H_4-folate, labeled with tritium exclusively in the 6-position hydrogen, have shown that the chick embryo synthetase also forms tritiated thymidylate (7); evidently an intramolecular transfer of a hydride ion is involved because no dilution of radioactivity occurs during introduction of the isotope into the methyl group.

According to the proposed mechanism, formation of the methylene bridge should also require the displacement of a hydrogen from carbon-5 of uracil, and Lomax and Greenberg (10) have demonstrated such displacement of the tritium atom of deoxyuridylate-5-³H during thymidylate synthesis catalyzed by the E. coli synthetase. The displaced tritium equilibrates with water and its release from the substrate is in direct proportion to the formation of product, thereby providing a highly sensitive assay for the enzymatic reaction.

Lomax and Greenberg (11) have also demonstrated that the E. coli synthetase will catalyze an exchange of hydrogen between water and the hydrogen at postion-5 of deoxyuridylate; in effect, this is the reverse of the process discussed in the preceding paragraph. This was shown by the incorporation of tritium into deoxyuridylate at position-5 when the thymidylate synthetase reaction was conducted in tritiated water with an excess of deoxyuridylate. The exchange reaction is interpreted as evidence that the formation of the methylene bridge between deoxyuridylate and $5,10$-methylene H_4-folate is reversible.

Proposed mechanisms for the thymidylate synthetase reaction are discussed in more detail by Blakley (2).

C. INHIBITION OF THYMIDYLATE FORMATION

The formation of thymidylate is specifically blocked by certain analogues of folic acid and by derivatives of 5-fluorouracil; these agents have an important use in the treatment of cancer in humans.

1. 5-Fluorouracil and 5-Fluorodeoxyuridine

The fluoropyrimidine antimetabolites have been reviewed by Heidelberger (12), who has contributed importantly to studies of the metabolism and biological effects of these compounds. 5-Fluorouracil (FU) and 5-fluorodeoxyuridine (FUdR) are clinically useful drugs, producing objective responses in the treatment of carcinoma of the breast and gastrointestinal tract. However, these very toxic chemicals also damage those normal tissues which have high rates of cell division, notably bone marrow and gastrointestinal epithelium. Fluorouracil and fluorodeoxyuridine are presumed to exert their principal toxic and therapeutic effects through inhibition of thymidylate synthetase and thereby, by blockade of DNA synthesis.

The anabolic metabolism of these compounds has been studied extensively in transplantable mouse tumor cells; it proceeds as follows:

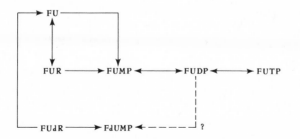

Both fluorouracil and fluorodeoxyuridine are converted to the deoxyribonucleoside monophosphate, fluorodeoxyuridylate (FdUMP); however, the di- and triphosphate derivatives are not formed. This probably explains why fluorouracil is not incorporated into DNA. Although fluorouracil is incorporated into RNA, with consequent biological effects, the specific and potent inhibition of thymidylate synthetase by FdUMP is held to account for the toxic and therapeutic effects of fluorouracil and fluorodeoxyuridine in animal cells. The formation of FdUMP in cells has been demonstrated, but the exact biochemical steps involved are not known.

Cohen *et al.* (*13*) discovered that fluorodeoxyuridylate (but not fluorouridylate) is a potent inhibitor of the phage-induced thymidylate synthetase in *E. coli* and is competitive with deoxyuridylate. Since this discovery, inhibition of the synthetase from other sources has been studied (see review, reference *2*). The inhibition by fluorodeoxyuridylate of the *E. coli* phage synthetase, or that from chick embryo or Ehrlich ascites cells, is competitive with respect to deoxyuridylate when all of the reaction components are added together. However, noncompetitive kinetics and a "stoichiometric" binding of the inhibitory nucleotide are obtained with the chick enzyme, if it is first incubated with fluorodeoxyuridylate and 5,10-methylene H_4-folate. Evidently the cofactor must be present for fluorodeoxyuridylate to bind in the noncompetitive manner (*7*). Similarly, it has been found that the bacterial enzyme can be virtually titrated with the inhibitor, if enzyme and inhibitor are preincubated prior to assay of the reaction. The Ehrlich cell enzyme differs in that fluorodeoxyuridylate is competitive with deoxyuridylate, in spite of preincubation with the enzyme.

It should be emphasized that fluorodeoxyuridylate is a very potent inhibitor of the synthetase; as a competitive inhibitor of deoxyuridylate in thymidylate synthesis by the enzyme from Ehrlich ascites tumor cells, the inhibition constant for fluorodeoxyuridylate is in the order of 10^{-8} M (*8*).

2. Folic Acid Antimetabolites

It was found in early studies that the folic acid analogues, methotrexate (amethopterin) and aminopterin very effectively blocked the incorporation of labeled deoxyuridine and of labeled formate into DNA thymine; however, the incorporation of thymidine was not blocked. It was apparent, therefore, that the analogues interfered with the introduction of the methyl group into thymine, a process known to involve H_4-folate. When it became established that the antifolic agents were exceedingly potent inhibitors of tetrahydrofolate dehydrogenase (see Chapter 5), the mechanism of their inhibition of DNA synthesis was apparent.

That the folate analogues do not interfere directly with the action of thymidylate synthetase, or with the action of the H_4-folate coenzyme, was clearly shown by McDougall and Blakley (*5*) in experiments which demonstrated that aminopterin did not block thymidylate synthesis from deoxyuridylate and serine, provided that *substrate* amounts of H_4-folate were available. However, if synthesis of thymidylate depended upon *catalytic* amounts of folate and an active folate reductase, aminopterin prevented thymidylate formation.

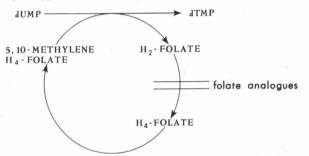

As noted in Chapter 5, aminopterin and methotrexate are bound very firmly to tetrahydrofolate dehydrogenase, with inhibition constants in the order of 10^{-9} M; the bound analogue can be released by dialysis against folic acid.

3. Inhibition of Cell Division

Both folate and uracil analogues exert their primary toxic effects on proliferating cells by preventing DNA synthesis through interruption of the synthesis of thymidylate; in other words, a "thymineless" state develops in cells exposed to these agents. It has been demonstrated many times with cultured cells that the amethopterin-induced inhibition of cell division may be released by exogenous thymidine, the provision of which results in a surge of DNA and mitotic activity in the cell population. Similarly, fluorodeoxyuridine produces a thymineless state, with a consequent inhibition of DNA synthesis and cell division; again, cells are released from this state by provision of exogenous thymidine. These effects have led to the use of amethopterin to synchronize cell populations with respect to DNA synthesis and cell division (*14*).

D. Deoxycytidylate Hydroxymethylase

The synthesis of 5-hydroxymethyldeoxycytidylate should be mentioned at this point, because this process is similar in many respects to the thymidylate synthetase reaction. The enzyme responsible, termed "deoxycytidylate hydroxymethylase," catalyzes an H_4-folate-dependent introduction of a hydroxymethyl group into the 5-position of deoxycytidylate.

Bacteriophages T2, T4, and T6 are unusual in that their DNA contains 5-hydroxymethylcytosine in place of cytosine; a large proportion of these hydroxymethyl groups are glucosylated. Shortly after infection of *E. coli* by these phages, the synthesis of bacterial DNA ceases and the synthesis of new, phage-specific DNA begins. The genome of the infecting phage brings into the bacterial cell information necessary for the production of a set of enzymes involved in the synthesis of the unusual DNA. The hydroxymethylase is one such phage-coded enzyme; a number of others are also known, including a specific kinase for hydroxymethyldeoxycytidylate, and a T2-DNA transglucosylase.

The hydroxymethylase catalyzes the following reaction, which is similar to that of thymidylate synthetase:

$$dCMP + 5,10\text{-methylene } H_4\text{-folate} \rightarrow 5\text{-hydroxymethyl-dCMP} + H_4\text{-folate}$$

The hydroxymethylase activity cannot be detected in uninfected *E. coli*. It may be noted that this reaction differs from thymidylate synthesis in that a reduction is *not* involved, as indicated by the fact that H_4-folate is a product. In studies of the mechanism of the reaction catalyzed by the the T4-induced hydroxymethylase, Yeh and Greenberg (*15*) have shown that tritium is displaced from deoxycytidylate-5-³H with the formation of product. As well, the enzyme catalyzed an H_4-folate-dependent exchange between water and the hydrogen atom at the 5-position of dCMP.

II. Formation of Deoxyuridylate

A. INTRODUCTION

As noted above, the substrate for thymidylate synthetase is deoxyuridylate, which may be derived endogenously in several ways and which may also be formed from exogenous deoxyuridine:

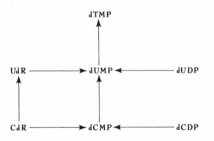

B. FROM EXOGENOUS DEOXYRIBONUCLEOSIDES

Deoxyribonucleosides are not obligatory intermediates in the endogenous formation of deoxyribonucleotides and may, therefore, be thought of as catabolites. Nevertheless, the enzymatic machinery for nucleoside anabolism exists and these compounds are useful to cells. Deoxyribonucleosides are found in appreciable concentrations in mouse and rat blood, and also appear in urine (16); deoxycytidine is present in rat blood plasma in concentrations of about 30 mμmoles/ml (17) and accounts for most of the deoxyribosidic material in rat blood (16).

As seen in the diagram above, both deoxycytidine and deoxyuridine may be converted to deoxyuridylate. Deamination of the former by cytidine deaminase yields deoxyuridine, which is phosphorylated by thymidine kinase (see Chapter 14). Cytidine deaminase is responsible for the deamination of both cytidine and deoxycytidine (see Chapter 12); the enzyme may be under regulatory control in that it is inhibited by dTTP (18).

C. From Deoxyuridine Di- and Triphosphates

The deoxyribonucleotides are derived primarily, if not entirely, by reduction of ribonucleoside phosphates; each of the several known ribonucleotide reductases (see Chapter 16) accepts as substrates phosphate esters of all four ribonucleosides, adenosine, guanosine, cytidine, and uridine. The ultimate fate of the uracil-containing reduction product, dUDP (or dUTP, in some microorganisms) is conversion to thymidine phosphates by way of deoxyuridylate and the thymidylate synthetase reaction.

In animal cells, the product of ribonucleotide reduction, dUDP, is converted to the two other levels of phosphorylation by reversible, ATP-dependent transphosphorylations catalyzed by thymidylate kinase (see below) and the general nucleoside diphosphate kinase.

In *E. coli*, as in animal cells, the direct product of the ribonucleotide reductase reaction is dUDP, which is converted to dUTP by kinase activity. As though to prevent possible incorporation of dUTP into DNA, a potent and specific pyrophosphatase is present in *E. coli;* this enzyme degrades dUTP to deoxyuridylate (19, 20). This activity is not known to exist in animal cells. These relationships are summarized in the following diagram:

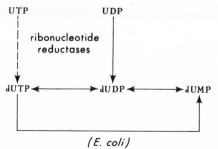

(E. coli)

D. FROM DEOXYCYTIDINE PHOSPHATES IN ANIMAL CELLS

The conversion of cytosine nucleosides to DNA thymidylate has been demonstrated repeatedly since the original work of Reichard and Estborn (21). The Maley and Maley (22) showed that labeled deoxycytidylate was converted to DNA thymidylate in a DNA-synthesizing system prepared from chick embryo and, thus, the existence of a deoxycytidylate deaminase was indicated. This enzyme, which is described below, and the cytidine–deoxycytidine deaminase activity are the only deaminases for cytosine derivatives known to occur in animal cells.

The ribonucleotide reductases form deoxycytidine phosphates which are destined for incorporation into DNA; these nucleotides are diverted to some extent into the deoxyuridylate pool, and thence into the thymine nucleotides, through the action of deoxycytidylate deaminase. This diversionary flow into the thymine pathway is regulated by the demand for the terminal products, dTTP and dCTP; the valve controlling this flow is deoxycytidylate deaminase, the activity of which is subject to allosteric regulation by dTTP and dCTP.

Deoxycytidylate deaminase was first isolated from sea urchin eggs by Scarano and has since been demonstrated in many animal tissues; the deaminases from spleen and chick embryo have been partly purified (23, 24). The enzyme requires that the substrate must be a 4-aminopyrimidine deoxyribonucleoside 5'-monophosphate, and will accept the following substituents at the pyrimidine 5-position: methyl, hydroxymethyl, fluoro, iodo, and bromo. Notably, the following are not deaminated: cytidine, deoxycytidine, cytidylate, dCDP, and dCTP. The concentration of this enzyme in cells appears to be related to their proliferative state in that the enzyme is found in growing tissues, but activities are low in adult tissues such as liver.

E. FROM DEOXYCYTIDINE PHOSPHATES IN BACTERIA

Deoxycytidylate deaminase is evidently absent from a number of bacterial species, including *E. coli*, *S. typhimurium*, and *B. subtilis*. Accordingly, in these cells the deoxycytidine phosphates do not contribute to the synthesis of the thymidine phosphates in the manner described above for animal cells. However, enzymes catalyzing the deamination of dCTP have been demonstrated in *E. coli* and *S. typhimurium*; the latter enzyme has been partly purified and shown to deaminate dCTP, but not CTP, dCDP, CDP, cytidylate, nor deoxycytidine (25). The formation of dUTP by this enzyme, followed by the action of the specific dUTP pyrophosphatase, would consitutue a route for the formation of deoxyuridylate:

$$dCTP \rightarrow dUTP \rightarrow dUMP$$

The work of Karlström and Larsson (*26*), discussed in Chapter 14 (see Table 14–II), illustrates the functioning of this route. Karlström and Larsson showed that a pyrimidine-requiring *E. coli* mutant (OK305), which was also deficient in deaminase activities for cytidine and deoxycytidine, synthesized DNA thymidylate from cytidine in a way which primarily involved the deoxycytidine phosphates. It is seen in Table 14–II that the labeling pattern of DNA thymidylate resembled that of DNA deoxycytidylate and was very different from that of RNA uridylate. Conversion of most of the incorporated cytidine to deoxyuridylate (and thence to the thymidine phosphates) by a route which did not involve uridine phosphates would appear to explain these data (*27*):

$$CR \to \to \to dCTP \to dUTP \to dUMP \to dTMP$$

F. From Deoxycytidine Phosphates in Phage-Infected Bacteria

As already mentioned, deoxycytidylate deaminase activity is not present in *E. coli* (*28*); however, on infection with bacteriophages T2, T4, and T6, the synthesis of the deoxycytidylate deaminase occurs. This and various other phage-specified enzymes enable the synthesis of the DNA peculiar to the T-even phages which contains hydroxymethylcytosine in place of cytosine. The T2-specified deoxycytidylate deaminase is regulated by nucleotides in much the same way as the animal cell deaminase, in that dTTP is inhibitory and dCTP activates (as well, 5 hydroxymethyl-dCTP activates the phage deaminase). It would appear that the phage-specified enzyme has the function of producing deoxyuridylate for eventual conversion of thymidylate of phage DNA. The specificity of the phage-specified deaminase is interesting in that 5-hydroxymethyldeoxycytidylate is not deaminated, although this compound is a substrate for the animal enzyme (*29*). The specificity of the phage deaminase would appear to preserve intracellular concentrations of 5-hydroxymethyldeoxycytidine phosphates destined for phage DNA.

Infection with phage T2 also induces the formation of an enzyme specific for the conversion of dCTP and dCDP to deoxycytidylate. T2 phage-specified enzymes catalyze the reactions shown in this diagram:

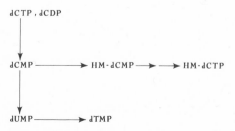

III. Deoxyribonucleotide Kinases

A. INTRODUCTION

The existence of kinases which convert deoxyribonucleoside monophosphates to their triphosphates has been recognized for a considerable time. For example, Kornberg and his co-workers (*30*), in their pioneering experiments with DNA polymerase, used enzymatically prepared triphosphates labeled in the α-phosphate; enzyme preparations partly purified from *E. coli* extracts were employed to convert α-^{32}P-labeled deoxynucleoside monophosphates to the α-labeled triphosphates.

Resolution of the enzymatic steps involved in converting each particular deoxyribonucleoside monophosphate to its di- and triphosphate derivatives has been accomplished through isolation and purification of the participating enzymes. It has turned out that in the deoxy series, β- and γ-phosphates are added in the same way as in the ribo series; in fact, it appears that most of the transphosphorylations are accomplished by kinases which will accept either the ribosyl or deoxyribosyl version of their particular substrates. Thus, the three levels of phosphorylation in the deoxyribonucleotide pool are interconnected by freely reversible transphosphorylation reactions (see also Chapter 4).

$$\text{BdRP} \overset{\text{ATP}}{\longleftrightarrow} \text{BdRPP} \overset{\text{NTP}}{\longleftrightarrow} \text{BdRPPP}$$

Phosphorylation of the monophosphates is effected by a *group* of enzymes which are specific for the bases; phosphorylation of the diphosphates is catalyzed by the ubiquitous nucleoside diphosphate kinase which has low specificity for both the base and pentose portions of its substrates. The latter enzyme occurs in animal cells in such high concentrations that the existence of the first step has been difficult to demonstrate in certain instances.

B. DEOXYRIBONUCLEOSIDE MONOPHOSPHATE KINASES

Fractionation of extracts from calf thymus for kinase activity toward deoxyribonucleoside monophosphates has revealed the presence of at least four separate enzymes; the kinase activities for dAMP, dGMP, dCMP, and dTMP are separate entities in this tissue (*31*). Each has been partly purified and examination of substrate specificities showed that the kinases for deoxyadenylate and deoxyguanylate also phosphorylate the ribosyl homologues, adenylate and guanylate, respectively. The deoxycytidylate kinase accepts as substrates both cytidylate and uridylate, but will not phosphorylate deoxyuridylate. The calf thymus thymidine monophosphate kinase was also isolated as a distinct entity; a very highly purified preparation of this enzyme from mouse hepatoma (*32*) accepts deoxyuridylate as a substrate (see further discussion below and in Chapter 4).

As a further instance of a dual function for the nucleoside monophosphate kinases, separate kinases for adenylate–deoxyadenylate, guanylate–deoxyguanylate, cytidylate–deoxycytidylate, uridylate, and thymidylate have been demonstrated in *E. coli* extracts (*33*); it should be noted that cytidylate-deoxycytidylate kinase and uridylate kinase were found as separate enzyme activities in the *E. coli* experiments.

Thus, it appears that a family of nucleoside monophosphokinases exists and that, although these enzymes have distinctive base specificities, they accept substrates of both the ribosyl or deoxyribosyl series.

Thymidine monophosphate kinase, which has a high specificity for the deoxyribosyl group, is discussed separately below.

C. Nucleoside Monophosphate Kinase Specified by Phage T2

Another instance of the induction of enzymes not ordinarily present in *E. coli* which occurs upon infection with bacteriophage T2 (see Section I, D and II, E) is found in the work of Bello and Bessman (*34*), who isolated from T2-infected *E. coli*, a deoxynucleoside monophosphate kinase with specificities quite distinct from those of the kinases discussed above. The phage-induced kinase is specific for *deoxy*ribonucleoside monophosphates and will phosphorylate three of the four nucleotides destined for phage DNA: deoxyguanylate, thymidylate, and hydroxymethyldeoxycytidylate. The enzyme does not phosphorylate deoxyadenylate.

D. Deoxyribonucleoside Diphosphate Kinase

The existence of a very active, widely distributed nucleoside diphosphate kinase has been demonstrated repeatedly (see Chapter 4 and reference *35*). This enzyme has been prepared in crystalline form from yeast (*36*), and Mourad and Parks (*37*) have isolated the enzyme in high purification from human erythrocytes. The latter enzyme reacts with nucleoside di- and triphosphates which contain either ribose or deoxyribose and any of the natural purine or pyrimidine bases, including thymine.

E. Synthesis of Thymidine Di- and Triphosphates

The conversion of thymidylate to the triphosphate in proliferating cells has received much attention because of the special position of thymidylate in the metabolic sequences leading to DNA. The conversion of thymidylate to dTTP has been used routinely as a measure of "thymidylate kinase" activity in unfractionated extracts; because of the presence of the ubiquitous and very active nucleoside diphosphate kinase, dTDP did not appear as an intermediate in most experimental systems. This led some to ask whether dTTP formation could be accomplished by pyrophosphorylation.

This question was resolved by demonstrations of the intermediate formation of dTDP by Grav and Smellie (38) and by Ives (39). Ives studied the phosphorylation of thymidylate in extracts of the Novikoff rat hepatoma, in which nucleoside diphosphate kinase activity was about three orders of magnitude higher than that of the other enzymes of thymine metabolism present (thymidine and thymidylate kinases, thymidylate synthetase, and thymidylate phosphatase). Ives showed that dTDP was an intermediate in dTTP formation by a method which was independent of the monophosphate and diphosphate kinase reactions. When ATP[γ^{32}P] was employed as the phosphorylating agent in the tumor extracts, the dTTP product was labeled in both γ- and β-phosphates; when ATP[β-^{32}P] was employed, dTTP was essentially unlabeled:

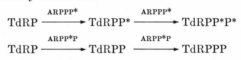

Thus, it was concluded that phosphorylation of thymidylate took place by the sequential addition of two monophosphate groups, rather than by transfer of a pyrophosphate group.

The enzyme responsible for the first phosphorylation is thymidine 5'-monophosphate kinase, called thymidylate kinase elsewhere; the second phosphorylation is accomplished by the general nucleoside diphosphate kinase.

F. THYMIDYLATE KINASE IN PROLIFERATING CELLS

The fact that thymidylate kinase activities are elevated in proliferating cells such as those of regenerating liver and tumors has prompted a good deal of experimentation related to the possibility that this enzyme might be important in the initiation of DNA synthesis. At the present time, nothing conclusive can be said about this possibility, but it will be recognized that a very impressive array of regulatory controls is focused on thymidylate metabolism and that dTTP has been implicated in the regulatory control of many enzymes of nucleotide metabolism.

A number of investigators have shown that thymidylate kinase is elevated in regenerating liver and that this increase coincides with, or just precedes, the onset of DNA synthesis (for example, see reference 40).

Synchronized cultures of HeLa cells display fluctuations in levels of thymidylate kinase activity that are related to stages of the mitotic cycle; thus, thymidylate kinase activity rises at the time of DNA synthesis and falls during division (*41*).

The enzyme does not appear to be regulated allosterically by nucleotides (*32*). Thus, regulatory influence by the kinase on the flow of metabolic traffic through the thymidine phosphates evidently would have to be exerted through changes in its concentration. In the latter connection, it may be relevant that thymidylate kinase is a very unstable activity and that a large portion of the mouse liver enzyme is contained in a latent, inactive form in mitochondria.

IV. Summary

Reaction sequences by which the pyrimidine deoxyribonucleosides and deoxyribonucleotides are interconverted are summarized below (steps

TABLE 15–I

ENZYMES OF DEOXYRIBONUCLEOTIDE INTERCONVERSION[a]

	Enzyme Commission number	Systematic name	Trivial name
1	3.5.4.5	Cytidine aminohydrolase	Cytidine deaminase
2	2.7.1.21	ATP:thymidine 5′-phospho-transferase	Thymidine kinase
3	—	—	Deoxycytidine kinase
4	—	—	Deoxycytidylate deaminase
5	—	—	Deoxycytidylate kinase
6	2.7.4.9	ATP:thymidinemonophosphate phosphotransferase	Thymidylate kinase
7	2.7.4.6	ATP:nucleosidediphosphate phosphotransferase	Nucleoside diphosphate kinase
8	—	—	Thymidylate synthetase
9	—	—	Deoxycytidylate hydroxy-methylase (T-even phage)
10	—	—	Ribonucleotide reductase
11	—	—	Deoxycytidine triphosphate deaminase
12	—	—	Deoxyuridine triphosphate pyrophosphatase
13	—	—	Deoxyribonucleoside mono-phosphate kinase (T2 phage)

[a] Numbers in first column refer to reactions in summary diagram (p. 242).

known only for bacteria or for T-even phages are shown with dashed lines). The enzymes are listed in Table 15–I.

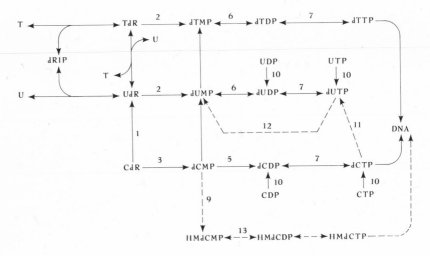

References

1. Crosbie, G. W., *in* "The Nucleic Acids" (E. Chargaff and J. N. Davidson, eds.), Vol. 3, p. 323. Academic Press, New York, 1960.
2. Blakley, R. L., *in* "The Biochemistry of Folic Acid and Related Pteridines" (A. Neuberger and E. L. Tatum, eds.), Vol. 13, Amer. Elsevier, New York, 1969.
3. Friedkin, M., and Kornberg, A., *in* "The Chemical Basis of Heredity" (W. D. McElroy and H. B. Glass, eds.), p. 609. Johns Hopkins Press, Baltimore, Maryland, 1967.
4. Huennekens, F. M., *Biochemistry* **2**, 151 (1963).
5. McDougall, B. M., and Blakley, R. L., *J. Biol. Chem.* **236**, 832 (1961).
6. Barner, H. D., and Cohen, S. S., *J. Biol. Chem.* **234**, 2987 (1959).
7. Lorenson, M. Y., Maley, G. F., and Maley, F., *J. Biol. Chem.* **242**, 3332 (1967).
8. Reyes, P., and Heidelberger, C., *Mol. Pharmacol.* **1**, 14 (1965).
9. Pastore, E. J., and Freidkin, M., *J. Biol. Chem.* **237**, 3802 (1962).
10. Lomax, M. I. S., and Greenberg, G. R., *J. Biol. Chem.* **242**, 109 (1967).
11. Lomax, M. I. S., and Greenberg, G. R., *J. Biol. Chem.* **242**, 1302 (1967).
12. Heidelberger, C., *Progr. Nucl. Acid Res. Mol. Biol.* **4**, 1 (1965).
13. Cohen, S. S., Flaks, J. G., Barner, H. D., Loeb, M. R., and Lichenstein, J., *Proc. Nat. Acad. Sci. U.S.* **44**, 1004 (1958).
14. Mueller, G. C., Kajiwara, K., Stubblefield, E., and Reuckert, R., *Cancer Res.* **22**, 1084 (1962).
15. Yeh, Y.-C., and Greenberg, G. R., *J. Biol. Chem.* **242**, 1307 (1967).
16. Rotherham, J., and Schneider, W. C., *J. Biol. Chem.* **232**, 853 (1958).
17. Guri, C. D., Swingle, K. F., and Cole, L. J., *Proc. Soc. Exp. Biol. Med.* **129**, 31 (1968).
18. Wisdom, G. B., and Orsi, B. A., *Eur. J. Biochem.* **7**, 223 (1967).

19. Bertani, L. E., Haggmark, A., and Reichard, P., *J. Biol. Chem.* **238**, 3407 (1963).
20. Greenberg, G. R., and Sommerville, R. L., *Proc. Nat. Acad. Sci. U.S.* **48**, 247 (1962).
21. Reichard, P., and Estborn, B., *J. Biol. Chem.* **188**, 839 (1951).
22. Maley, G. F., and Maley, F., *J. Biol. Chem.* **236**, 1806 (1961).
23. Scarano, E., Geraci, G., Polzella, A., and Companile, E., *J. Biol. Chem.* **238**, PC1556 (1963).
24. Maley, G. F., and Maley, F., *J. Biol. Chem.* **239**, 1168 (1964).
25. O'Donovan, G. A., and Neuhard, J., *Bacteriol. Rev.* **34**, 278 (1970).
26. Karlström, O., and Larsson, A., *Eur. J. Biochem.* **3**, 164 (1967).
27. Karlström, O., personal communication (1971).
28. Keck, K., Mahler, H. R., and Fraser, D., *Arch. Biochem. Biophys.* **86**, 85 (1960).
29. Maley, G. F., and Maley, F., *J. Biol. Chem.* **241**, 2176 (1966).
30. Lehman, I. R., Bessman, M. J., Simms, E. S., and Kornberg, E., *J. Biol. Chem.* **233**, 163 (1958).
31. Sugino, Y., Teraoka, H., and Shimono, H., *J. Biol. Chem.* **241**, 961 (1966).
32. Kielley, R. K., *J. Biol. Chem.* **245**, 4204 (1970).
33. Hiraga, S., and Sugino, Y., *Biochim. Biophys. Acta* **114**, 416 (1966).
34. Bello, L. J., and Bessman, M. J., *J. Biol. Chem.* **238**, 1777 (1963).
35. Weaver, R. H., *in* "The Enzymes" (P. D. Boyer, H. Lardy, and K. Myrbäck, eds.), 2nd rev. ed., Vol. 6, p. 151. Academic Press, New York, 1962.
36. Ratliff, R. L., Weaver, R. H., Lardy, H. A., and Kuby, S. A., *J. Biol. Chem.* **239**, 301 (1964).
37. Mourad, N., and Parks, R. E., Jr., *J. Biol. Chem.* **241**, 271 (1966).
38. Grav, H. J., and Smellie, R. M. S., *Biochem. J.* **89**, 486 (1963).
39. Ives, D. H., *J. Biol. Chem.* **240**, 819 (1965).
40. Fausto, N., and Van Lancker, J. L., *J. Biol. Chem.* **240**, 1247 (1965).
41. Brent, T. P., Butler, J. A. V., and Crathorn, A. R., *Nature (London)* **207**, 176 (1965).

CHAPTER 16

ENZYMATIC REDUCTION OF RIBONUCLEOTIDES

I. Introduction

In nature, the deoxyribose constituent of deoxyribonucleotidic materials, free or polymerized, originates from the enzymatic reduction of ribonucleotides. Although enzymatic machinery exists for the incorporation or reutilization of preexisting deoxyribosyl compounds, they have their primary origins in the reduction of ribonucleotides. An alternative possibility, that of a contribution to deoxyribosyl synthesis through the action of deoxyribo-aldolase, has been mentioned previously; the weight of evidence presently at hand indicates that this enzyme should be thought of as having a catabolic function.

The conversion of ribose to deoxyribose takes place at the free nucleotide level. The essential step in this process is the replacement of the 2'-hydroxyl of ribonucleotides by a hydrogen atom. By introduction of a hydride ion the original hydroxyl group is displaced without changes in other substituents of the ribofuranose ring.

A. Evidence That Ribosyl Reduction is a Source of the Deoxyribosyl Moiety in DNA

The concept that the deoxyribosyl groups of DNA nucleotides are derived by reduction of ribonucleoside derivatives is based on a wide array of experimental evidence.

In 1950, Hammarsten, Reichard, and Saluste (1) commented on the likelihood of a ribosyl to deoxyribosyl conversion, perceiving the existence of this route from the observation that although free pyrimidine bases were poorly incorporated into DNA in animal tissues, pyrimidines in the form of *ribo*nucleosides were incorporated into DNA many times more efficiently.

Subsequently, others measured the conversion of isotopically labeled ribonucleosides into DNA nucleotides in experiments which showed that this process took place without cleavage of the N-glycosidic bond. In such experiments, the test nucleosides contained isotopic label in both sugar and base. After administration of the doubly labeled nucleoside to cells, both DNA and RNA were isolated and degraded to nucleotides; the relationships between the isotope content of the base and sugar portions of the isolated deoxyribonucleotides were compared with those of the appropriate ribonucleotides from RNA. Interpretation of this type of experiment was conceived as follows: if the relationship between base and sugar labeling in a deoxyribonucleotide from DNA was found to be the same as that for the corresponding RNA-derived ribonucleotide, then no isotopic dilution would have occurred in the sugar relative to the base during ribosyl reduction. Such a result would show that (a) the reduction had not involved as intermediates free ribosyl compounds which had equilibrated with the substantial cellular pools of ribosyl phosphates and (b) deoxyribosyl synthesis from ribosyl precursors took place without the intermediate cleavage of the base–sugar linkage.

In experiments of this type with animal tissues, Rose and Schweigert (2) and Thomson et al. (3) showed that pyrimidine ribonucleosides were converted to DNA nucleotides without cleavage of the N-glycosidic bond. As well, Larsson and Neilands (4) performed a similar type of experiment in which ^{32}P-phosphate and uniformly labeled ^{14}C-cytidine were administered to rats with regenerating liver. Both substances were incorporated into the liver polynucleotides which, upon isolation, were degraded to their constituent nucleotides for analysis. Their data (Table 16–I) showed that the four nucleotides of RNA had similar specific activities with respect to ^{32}P, indicating that the labeled phosphate readily equilibrated with the nucleoside phosphate pool during the experimental period. In this experiment, cytidylate derived from RNA and deoxycytidylate derived from DNA had the same ^{32}P:^{14}C ratio. This result indicated that both polynucleotide subunits, deoxycytidylate and cytidylate, were derived from a common precursor, evidently a ribonucleotide.

The experiments of Karlström and Larsson referred to in Chapter 14 have shown very clearly that in $E.$ $coli$, cytidine can be converted into RNA cytidylate and DNA deoxycytidylate without dilution of label in the sugar moiety relative to the base; the data of Table 14-II illustrate

TABLE 16–I

DISTRIBUTION OF ^{32}P AND ^{14}C IN THE NUCLEOTIDES OF REGENERATING RAT LIVER RNA AND DNA AFTER INJECTION OF NaH$_2$^{32}PO$_4$ AND ^{14}C-CYTIDINE[a]

| Nucleotide | Specific activity (cpm per μmole) | | Ratio |
	^{32}P	^{14}C	^{32}P/^{14}C
CMP	13,170	5,960	2.24
AMP	16,600	nil	—
UMP	16,650	290	—
GMP	12,800	nil	—
dCMP	5,100	2,220	2.29
dAMP	9,750	nil	—
dTMP	8,650	200	—
dGMP	6,880	nil	—

[a] From (4). Reproduced with permission.

this point. These experiments show clearly that the deoxyribosyl portion of deoxycytidylate was derived by the *direct* reduction of a cytidine metabolite.

B. THE ENZYMATIC BASIS OF RIBOSYL REDUCTION

Following the above-mentioned demonstration by Rose and Schweigert (*2*) that deoxyribosyl groups were derived by direct reduction of ribosyl groups, efforts were made to characterize this process in biochemical and enzymological terms. Grossman and Hawkins (*5*) were the first to develop a cell-free system capable of the reduction; they showed that extracts of *S. typhimurium* formed deoxycytidine compounds from cytidine and found this process to be enhanced by thiols. With cell-free extracts from chick embryos, Reichard and his collaborators (see review, reference *6*) demonstrated that each of the four ribonucleotides found in RNA were reduced; although ribonucleo*tides* were involved, it was not apparent whether the participating compounds were at the mono-, di-, or triphosphate level. For technical reasons, the Reichard group then turned to extracts of *E. coli* and established that, in this system, the substrates and products of the reaction were nucleoside diphosphates. This discovery provided the critical focus required for the biochemical dissection of the reduction system, which was then undertaken. Reichard and co-workers have since isolated and purified the four protein elements of the *E. coli* system for ribonucleotide reduction. Ribonucleotide reductase was found to be comprised of two subunits, proteins B1 and B2, which individually were inactive. The reducing power for ribonucleotide reduction was found to be provided by a hydrogen transport system consisting of two components, thioredoxin and a flavoprotein, thioredoxin reductase; the latter is NADP-dependent.

The discoveries made with the *E. coli* system provided the basis for studies of ribonucleotide reduction in other microbial species and in animal cells. The mechanism of ribonucleotide reduction in rat tissues resembles that of *E. coli* in many ways, whereas ribonucleotide reduction in *Lactobacillus leichmannii* differs distinctively, in that coenzyme B_{12} takes part in a reduction accomplished at the nucleoside triphosphate level. Each of the three reductases that have been extensively purified reduces ribonucleotide substrates representing all four of the ribonucleosides of RNA and each displays remarkable allosteric regulatory properties (see below).

II. Ribonucleotide Reduction in *Escherichia coli*

A. FRACTIONATION OF THE RIBONUCLEOTIDE REDUCING SYSTEM

Reduction of cytidylate in crude extracts of *E. coli* required ATP and NADP; the discovery that a chemical reductant, dihydrolipoate, would replace the need for a NADP-dependent reducing system (which would likely involve one or more proteins) provided an assay which directly measured ribonucleotide reductase activity and, thereby, simplified fractionation of the bacterial extracts. The cytidylate-reducing system was then divided into a kinase-containing fraction and a ribonucleotide reductase fraction (7), and it was apparent that the reduction was taking place at the diphosphate level.

$$CMP \xrightarrow[\text{ATP, Mg}^{2+}]{\text{kinase}} CDP$$

$$CDP + \text{lipoate-(SH)}_2 \xrightarrow[\text{B1·B2}]{} dCDP + \text{lipoate-S}_2$$

As purification of the ribonucleotide reductase progressed, it became evident that the activity was present in a complex of two protein components, B1 and B2; these are now known to be nonidentical subunits of a remarkable allosteric enzyme. It was recognized that dihydrolipoate was unlikely to be the physiological reductant in this system.

The isolated reductase (above) was used in assays to direct further fractionation experiments which culminated in the isolation of the physiological reducing system; this turned out to be a previously unrecognized hydrogen transport system. This system, which connects ribonucleotide reductase to the NADPH-NADP+ system, was found to be a two-component system consisting of a small sulfhydryl protein, thioredoxin, and a flavoprotein, thioredoxin reductase. Thioredoxin is the reductant which specifically interacts with the ribonucleotide reductase. In the presence of catalytic amounts of thioredoxin, the thioredoxin reductase will link NADPH with the reduction of ribonucleoside diphosphates as follows:

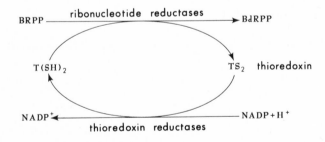

B. Ribonucleotide Reductase

During purification, the ribonucleotide reductase from *E. coli* separates into two fractions, B1 and B2. These fractions have been purified to virtual homogeniety; separately they are catalytically inactive, but together, in the presence of Mg^{2+}, a catalytically active complex is formed consisting of one molecule of each component. Formation of the B1–B2 complex has been demonstrated by Brown and Reichard (*8*) using sucrose density gradient centrifugation:

$$\text{B1 (7.8 S)} + \text{B2 (5.5 S)} \xrightarrow{\text{Mg}^{2+}} \text{B1–B2 (9.7 S)}$$

The reductase (B1–B2 complex) is an allosteric enzyme; both the activity and the substrate specificity of the reductase are modulated by nucleoside triphosphates. ADP, GDP, CDP, and UDP are all substrates for the reductase; however, the specificity of the enzyme toward these compounds is determined by allosteric effectors. In other words, the preference of the reductase toward these substrates depends upon which of the several nucleoside triphosphate effectors occupy the regulatory sites on the enzyme complex (see below). The B1 subunit has sites to which the regulatory nucleotides are bound, but the B2 subunit lacks such sites (*9*).

The B2 protein contains 2 moles of non-heme iron per molecule of enzyme. The iron is bound tightly, but can be removed by dialysis against 8-hydroxyquinoline. The iron-depleted enzyme is inactive, but activity may be restored by addition of iron. The B2 protein has a characteristic absorption peak at 410 mμ, which is attributed to its iron content. Hydroxyurea, which is known to inhibit DNA biosynthesis, evidently achieves this effect through interaction with the B2 protein (*7*). When the B2 protein is treated with hydroxyurea, the characteristic absorption at 410 mμ disappears, and the resulting loss of enzyme activity parallels the decrease in absorbance at that wavelength. Similar effects are obtained with hydroxylamine or hydrazine.

The *E. coli* ribonucleotide reductase has no requirement for coenzyme B$_{12}$. It became a matter of some importance to establish this point, because the ribonucleotide reductase of *L. leichmannii* has an absolute requirement for this coenzyme, which it binds very tightly. That the *E. coli* reductase does not bind B$_{12}$ was shown by experiments in which *E. coli* was grown in the presence of ^{60}Co-vitamin B$_{12}$; the reductase isolated from such cells contained no radioactivity (*10*).

Synthesis of the *E. coli* reductase is under repressor control and thymine derivatives are the determining metabolites. This was shown by Beck and co-workers (*11*) with *E. coli*$_{15T^-}$, a thymine-requiring mutant which lacks thymidylate synthetase. When log phase cells of this mutant were transferred to a thymine-deficient medium, a tenfold increase in the activity of the reductase resulted. The use of such derepressed cells greatly facilitated the purification of the reductase. Synthesis of the reductase in wild-type *E. coli* is also derepressed in the presence of fluorodeoxyuridine, which induces a thymine deficiency through inhibition of thymidylate synthetase.

As well, the synthesis of the ribonucleotide reductase of *L. leichmannii* is under repressor control.

C. The Thioredoxin Hydrogen Transport System

The ribosyl–deoxyribosyl transformation in *E. coli* and in other microbial systems employs the reducing potential of NADPH and is linked to the latter through a hydrogen transport system comprised of thioredoxin and thioredoxin reductase. It is likely that a similar system operates in animal cells; however, the components have not yet been highly purified. In *E. coli*, reduced thioredoxin (a dithiol) interacts with the B1–B2 complex, providing hydrogen atoms and electrons for the reduction; although the reductase will utilize certain other sulfhydryl compounds as reductants (for example, dihydrolipoate or dithioerythritol), thioredoxin is the natural reductant. Oxidized thioredoxin (a disulfide) is reduced by the NADPH-linked thioredoxin reductase and, thus, thioredoxin serves as a hydrogen-carrying shuttle between the two reductases.

1. *Thioredoxin*

The thioredoxins are small, heat-stable proteins; the *E. coli* thioredoxin has been purified to homogeneity and its complete pr.mary structure has been determined (*12*). Reduced thioredoxin is a single polypeptide chain of 108 amino acids (molecular weight 11,657) and contains two tryptophan and two cysteine residues. The N-terminal and the C-terminal amino acids are serine and alanine, respectively; the amino acid sequence in the vicinity of the cysteines is as follows:

$$...\,\text{Asp-Phe-Try-Ala-Glu-Try-Cys-Gly-Pro-Cys-Lys}\,...$$
$$\qquad\qquad\qquad\quad\underset{\text{SH}}{|}\qquad\qquad\underset{\text{SH}}{|}$$

The sulfhydryl form is oxidized by participation in ribonucleotide reduction with the formation of a disulfide bridge. This oxidation is accompanied by a conformational change in the molecule localized in the vicinity of the two tryptophan residues, both of which are located close to the sulfhydryl-bearing cysteines (see above). This conclusion was reached from a spectrofluorimetric study of the reduced and oxidized forms of thioredoxin which showed that reduction led to a large increase in the quantum yield of the tryptophan emission of thioredoxin; the conformational changes so indicated were evidently local and, therefore, near the active site of the protein, because other measurements indicated that no gross conformational changes had occurred upon reduction (*12*). It has been suggested that the conformational difference between the oxidized and reduced forms of thioredoxin may facilitate binding of thioredoxin molecules to whichever of the two reductases is appropriate to their oxidation state (for example, the oxidized form would bind to thioredoxin reductase).

In a general sense, reduced thioredoxins appear to be capable of serving as reductants to heterologous ribonucleotide reductases. For example, the *E. coli* thioredoxin will function as a hydrogen donor for the ribonucleotide reductase of the Novikoff hepatoma, and yeast thioredoxin will serve the *E. coli* reductase. On the other hand, reduction of any particular thioredoxin is catalyzed only by the homologous thioredoxin reductase, that is, both must be obtained from the same organism.

2. *Thioredoxin Reductase*

This flavoprotein catalyzes the reduction of oxidized thioredoxin by NADPH, according to the following reaction:

$$\text{Thioredoxin-S}_2 + \text{NADPH} + \text{H}^+ \longleftrightarrow \text{thioredoxin-(SH)}_2 + \text{NADP}^+$$

The *E. coli* enzyme has been purified to homogeneity; the molecular weight is about 66,000 and the enzyme consists of two identical or very similar polypeptide chains. The amino acid composition of the enzyme has been determined (*13*). It has been proposed that each of the two thioredoxin reductase subunits has an active center containing one FAD molecule and one disulfide linkage, both of which act as oxidation–reduction acceptors during catalysis (*13*). As well, thioredoxin reductase contains four additional sulfhydryl groups per molecule (presumably, two per subunit), but these do not participate in the catalytic function. The amino acid sequence at the active site of the *E. coli* thioredoxin reductase has been determined to be as follows (*14*):

$$\ldots \text{-Ala-Cys-Ala-Thr-Cys-Asp-Gly-Phe-} \ldots$$
$$\qquad\quad \overset{|}{\text{S}}\text{————}\overset{|}{\text{S}}$$

It will be noted that in the active centers of both thioredoxin (see above) and its reductase, the two cysteine residues are separated by only two amino acid residues.

3. *The Thioredoxin–Thioredoxin Reductase System*

Thioredoxin hydrogen transport systems have been demonstrated in various organisms other than *E. coli*, for example, in tumor cells, yeast, and *L. leichmannii*. As well, a thioredoxin is specified by the bacteriophage T4 genome. In general, reduced thioredoxins will serve as hydrogen donors in nucleotide reductions catalyzed by heterologous ribonucleotide reductases, that is, by reductases obtained from a different organism. On the other hand, thioredoxin reductases are specific and appear to reduce only the homologous thioredoxin. In this connection, it may be noted that thioredoxins from different organisms differ considerably in their amino acid composition. Bacteriophage T4 provides an exception to these generalizations. The phage genome specifies both a ribonucleotide reductase and a thioredoxin; the reductase appears to be specific for the phage thioredoxin in that it will not accept *E. coli* thioredoxin as a hydrogen donor (*15*). However, the phage thioredoxin is reduced by *E. coli* thioredoxin reductase and is as good a substrate for this enzyme as its ordinary substrate, the bacterial thioredoxin. These specificities appear to be intimately involved in the phage's parasitism of the host cell metabolism.

Reichard has noted that, although thioredoxin reductase is specific for thioredoxin, the coupled action of these two proteins may be able to serve as an NADPH-using hydrogen carrier in enzymatic reductions other than those of ribonucleotides (*7*). In agreement with this idea, it has been noted that adult rat liver contains a hydrogen transport system capable of serving ribonucleotide reductase, even though the tissue does not contain demonstrable levels of the latter enzyme (*16*).

D. Reaction Mechanism of Ribonucleotide Reduction

It is apparent from the foregoing that hydrogen borne by the sulfhydryls of thioredoxin is the "reagent" used in the enzymatic reduction of the ribosyl group. In this reaction, the 2'-hydroxyl group is displaced and its position occupied by an incoming hydrogen atom (probably in the form of a hydride ion, H^-). The other atoms of the ribosyl group remain in their original positions, including the 2'-hydrogen atom *cis* to the *N*-glycosidic bond. This retention of the steric arrangement at the 2'-position became apparent in experiments with purified components of the *E. coli* system, in which the reduction of CDP was conducted in water in which hydrogen was replaced by its isotopes, tritium or deuterium. Sulfhydryl hydrogens in general, and those of thioredoxin-$(SH)_2$, exchange very rapidly with water protons and, thus, become tritiated, or deuterated, by exchange when in solution in ^3H-water or ^2H-water, respectively. Larsson (*17*) showed that when the CDP reducing system contained ^3H-water, ^3H was transferred to the product, dCDP, and was fixed into the 2'-position. A degradation procedure showed this to be the exclusive location of the isotope in the product. A considerable isotope effect was apparent in this experiment and, although the position of the incoming isotope was clearly indicated, the *number* of hydrogen atoms introduced during the reduction could not be determined.

Similar experiments in which CDP was reduced in 99.7% D_2O, showed that one atom of deuterium was introduced stereospecifically into position-2'; nuclear magnetic resonance was the analytical technique employed (*18*). Interpretation of NMR spectra permitted the conclusion that the deuterium atoms were introduced on the same side of the pentose molecule as the original hydroxyl was located; this result indicated that the incoming hydrogen in the reduction does not induce changes in configuration at C-2. This feature of the reduction eliminated a number of possible reaction mechanisms, including the formation of a double bond between C-2 and C-3. Similar experiments have shown that deuterium from D_2O is incorporated into the 2'-position of dATP during the reduction of ATP by the *L. leichmannii* reductase (*19*). Thus, it appears that the mechanism by which the 2'-hydroxyl is replaced by hydrogen is similar in both bacterial reductase systems, even though their components are quite different.

These results have been explained by a mechanism which supposes that a hydride ion (H^-) is generated and that a concerted reaction takes place in which the leaving group (OH^-) dissociates from C-2, but only when the attacking species (H^-) is present (*7*, *20*). Reichard has suggested that the iron-containing B2 subunit of the *E. coli* ribonucleotide reductase may somehow provide the postulated hydride ion. In ribonucleotide reduction by the *L. leichmannii* system, 5'-deoxyadenosylcobalamin participates as a coenzyme and may also be involved in generating a hydride ion.

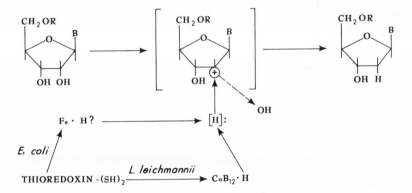

E. REGULATION OF THE RIBONUCLEOTIDE REDUCTASE

The ribonucleotide reductase of *E. coli* is an allosteric enzyme, the activity and specificity of which are modulated in a very complex manner by several nucleoside triphosphates. Both the Michaelis constant and V_{max} values for the individual substrates were found to be influenced by such effectors; when combinations of effectors were employed, the resulting patterns of activation and inhibition indicated that regulation of the catalytic activity was very complicated, indeed.

It became apparent that the substrate specificity of the reductase was determined by nucleoside triphosphate effectors. For example, the reductase purified in respect to CDP reduction, was only slightly active toward *purine* nucleoside diphosphates, even in the presence of ATP, a stimulator of CDP reduction. However, the reductase became active toward purine nucleoside diphosphates in the presence of dTTP, as illustrated in Fig. 16-1. It is seen in these data that dGTP and dTTP are potent positive effectors, greatly increasing the activity of the enzyme in the reduction of GDP. dATP is strongly inhibitory in this reaction, even blocking the promoting effect of dTTP; as will be evident subsequently, dATP is also a negative effector with other substrates.

In an experiment which is summarized in Fig. 16-2, Larsson and Reichard (*21*) showed that the specificity of the catalytic site in the enzyme may be changed by a nucleotide effector. These data show the influence of CDP concentration on the rate of its reduction and were determined in the presence of different concentrations of GDP. dTTP was present in Experiment A (left panel); ATP was the effector in Experiment B (right panel). In the presence of dTTP, GDP was a competitive inhibitor of CDP reduction; however, when ATP was substituted for the dTTP effector,

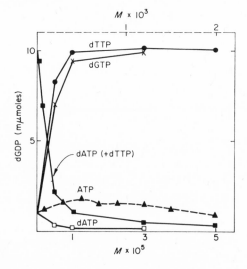

FIG. 16-1. Influence of allosteric effectors on reduction of GDP by ribonucleotide reductase from *E. coli*. From (*6*). Reproduced with permission.

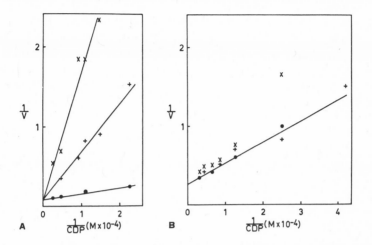

FIG. 16-2. Competition between GDP and CDP for *E. coli* ribonucleotide reductase. Increasing amounts of CDP were incubated without GDP ($\bullet$——$\bullet$), with $4 \times 10^{-5} M$ GDP (+———+), and with $1 \times 10^{-4} M$ GDP ($\times$———$\times$); A, with dTTP ($4 \times 10^{-5} M$) as allosteric effector; B, with ATP ($4 \times 10^{-4} M$) as allosteric effector. From (*21*). Reproduced with permission.

GDP did not compete in the reduction of CDP. In other words, the allosteric effector in this experiment, dTTP or ATP, determined whether or not the active site of the enzyme would accept GDP as a competitive substrate. For a much fuller exposition of these ideas, presented in the perspective of their historical development, the reader is referred to the review by Reichard (*7*).

The ability of nucleoside triphosphates, singly, or in combination, to effect changes in the catalytic properties of the reductase have been explored systematically by Reichard and his co-workers; the complicated pattern of results is summarized in Table 16–II. These findings have been interpreted in terms of four states of activity for the reductase. In the presence of ATP, the reductase prefers as substrates pyrimidine nucleoside diphosphates. In the presence of dGTP, a purine-specific state is assumed and dTTP causes the enzyme to be in a condition in which both purine and pyrimidine ribonucleoside diphosphates are reduced.

The effects of dATP are complex. Low concentrations of dATP stimulate reduction of pyrimidine ribonucleoside diphosphates, whereas at high concentrations, dATP becomes a potent, general inhibitor of nucleotide reduction; it is also seen that low concentrations of dATP counteract the stimulatory effects of dTTP and dGTP. It has been concluded that dATP converts the enzyme into an inactive state.

TABLE 16–II

INFLUENCE OF DIFFERENT EFFECTORS ON THE ACTIVITY AND SPECIFICITY
OF RIBONUCLEOTIDE REDUCTASE[a]

Nucleotide (M)	Catalytic activity	Specificity for base
ATP (2×10^{-3})	Stimulation	Pyrimidines
dATP (10^{-6})	Stimulation	Pyrimidines
dATP (10^{-4})	Inhibition	Purines + pyrimidines
dGTP (10^{-5})	Stimulation	Purines
dTTP (10^{-5})	Stimulation	Purines + pyrimidines
dATP (10^{-4}) + ATP (2×10^{-3})	Stimulation	Pyrimidines
dATP (10^{-6}) + dTTP (10^{-4})	Inhibition	Purines + pyrimidines
dATP (10^{-6}) + dGTP (10^{-4})	Inhibition	Purines + pyrimidines
ATP (10^{-4}) + dTTP (5×10^{-4})	Inhibition	(Purines) + pyrimidines

[a] From (9). Reproduced with permission.

The nucleoside triphosphate effectors themselves do not participate in the reaction; that is, they are unchanged by the reduction reaction, in spite of influences upon it. Their effects on the specificity of the reductase are exerted by both decreasing the values of Michaelis constants for substrates and by increasing maximum velocities in their reduction.

Recent publications by Brown and Reichard (8, 9) have related these effects on the catalytic activity of the *E. coli* reductase to physical and chemical properties of the individual subunits of the enzyme. These properties will first be considered in a general way. Brown and Reichard (9) demonstrated in dialysis studies that the B1 subunit of the reductase binds the nucleoside triphosphate effectors, whereas the iron-containing B2 subunit is inactive in this respect. The B1 protein has a molecular weight of about 200,000 and is evidently a dimer. The dialysis studies indicated that each B1 molecule possessed four nucleotide binding sites: two were termed "h-sites" because of their high affinity for dATP; the other two sites ("l-sites") bind dATP with a tenfold lower affinity than the h-sites. The h-sites also bind ATP, dGTP, and dTTP, and they are believed to be concerned with the *specificity* of the reductase. In addition to binding dATP, the l-sites also bind ATP, but do not bind dGTP or dTTP; the l-sites evidently influence the *activity* of the reductase. The interaction of dATP with the l-sites converts the reductase to an inactive state.

The active form of the *E. coli* reductase is the B1–B2 complex comprised of one molecule of each subunit. Effector binding to the B1 subunit of the B1–B2 complex is not changed in a qualitative sense from that found for the B1 portion alone. The B1–B2 molecule forms an inactive dimer in the presence of dATP, and this appears to be the basis of the general inhibitory action of dATP.

The effector binding data and the effector-induced changes in catalytic activity of the reductase have been correlated in a general scheme (9), which is presented in Fig. 16-3.

Figure 16-3 depicts only protein B1. The effector binding sites of the h-type (high affinity for dATP) are shown as solid circles; the l-sites are shown as solid squares. Illustrated are three variations on the pyrimidine-specific state of the enzyme; these states were apparent in experiments with the effectors indicated (see Table 16–II). These forms of the B1 subunits and that of the purine–pyrimidine specific state (effector: dTTP), are all about equally active in the reduction of CDP.

Three states of the enzyme with activity toward purine substrates are shown; the nucleotide effectors impart very clear differences in specificity toward the purine substrates. The dGTP-activated enzyme prefers ADP as a substrate, whereas the dTTP-activated enzyme prefers GDP.

Inhibition by dATP is visualized as a consequence of the binding of this effector to the l-site; an exception to this pattern is seen in the inactive state which results from the dTTP–ATP combination. In each case, the inactive state is characterized by the formation of a "heavy" enzyme complex, which is evidently a dimer of the B1–B2.

One can imagine, as Larsson and Reichard (22) have done, that the regulatory properties perceived through *in vitro* experiments, might operate in the living cell as a physiological mechanism which adjusts or balances production of the four deoxyribonucleotide building blocks needed for

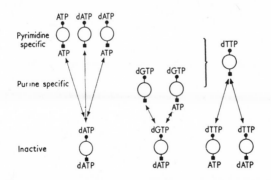

FIG. 16-3. Scheme for different forms of protein B1 of *E. coli* ribonucleotide reductase. The h-sites are represented by solid circles and the l-sites by solid squares. For clarity, only one site of each class is depicted. The arrows show transitions between active and inactive forms. From (9). Reproduced with permission.

DNA synthesis in order to avoid surpluses or shortages. It will be appreciated that in living cells there must be highly complicated interactions between the reductase and the ultimate products of the reduction, all of which may be present at once in a cell.

III. Ribonucleotide Reduction in Other Cells

As we have seen, the biosynthesis of deoxyribonucleotides in *E. coli* has been studied in considerable depth; discoveries made with that system greatly facilitated the exploration of this process in other cell types. Ribonucleotide reduction in several kinds of animal cells is evidently accomplished by a mechanism similar to that found in *E. coli*, whereas in *Lactobacillus leichmannii* and in certain *Rhizobium* and *Clostridium* species (*23*), the process of ribonucleotide reduction differs distinctively from that in *E. coli* in that B$_{12}$ cofactors are involved.

A. A B$_{12}$ Coenzyme-Requiring System in *L. leichmannii* and Other Microorganisms

Apart from that of *E. coli*, the only microbial system for deoxyribonucleotide synthesis which has been studied in detail is that of *L. leichmannii*. Prior to the definitive biochemical studies described below, nutritional experiments had demonstrated that the vitamin B$_{12}$ requirement of *L. leichmannii* was involved in an essential way in the biosynthesis of deoxyribonucleotides. Blakley and Barker (*24*) showed that cell-free extracts of this microorganism catalyzed the reduction of cytidylate to deoxycytidylate and showed also that this reaction required a vitamin B$_{12}$ derivative and NADP.

Synthesis of the *L. leichmannii* reductase is derepressed under particular culture conditions in which minimal amounts of vitamin B$_{12}$ derivatives are provided. Under such conditions, the aporeductase is produced in amounts which greatly exceed the amount of the B$_{12}$ cofactor present in the bacterial cells. Purification procedures developed for the reductase utilized cells derepressed in this way and yielded the highly purified apoenzyme (*25*); the holoenzyme was not isolated in these procedures. The isolation of the inactive aporeductase made possible the demonstration of its absolute requirement for a vitamin B$_{12}$ cofactor; 5′-deoxyadenosylcobalamin, the most abundant corrinoid in *L. leichmannii* (*26*), binds very tightly to the aporeductase and, therefore, appears to be the natural cofactor. The highly purified reductase accepts as substrates only ribonucleoside *tri*phosphates, in contrast to the requirement of the *E. coli* enzyme for diphosphates. A single enzyme reduces ATP, GTP, UTP, or CTP; the specificity of the enzyme toward the individual substrates is determined by nucleoside triphosphate effectors.

In the reduction of ribonucleoside triphosphates, the *L. leichmannii* enzyme is able to use as reductant, dihydrolipoate, dithiothreitol, or dithioerythritol, the oxidized forms of which are cyclic disulfides; it has been shown that for each mole of dihydrolipoate oxidized, a mole of ribonucleoside triphosphate is reduced. The physiological reductant for the *L. leichmannii* reductase is a two-component, hydrogen transport system analogous to the thioredoxin–thioredoxin reductase system of *E. coli*; the two components of the *L. leichmannii* system have been purified and are generally similar to those of the *E. coli* system, although their molecular weights are distinctive (*27*). The thioredoxin-thioredoxin reductase pair from *E. coli* will serve the *L. leichmannii* reductase, although the converse is not true.

As already mentioned, the reaction mechanism for ribonucleotide reduction by the *L. leichmannii* enzyme is analogous to the *E. coli* mechanism (see Section II, D, in this chapter); when the reduction was conducted in deuterated or tritiated water the isotope was incorporated only into the 2′-position of the deoxyribosyl group of the product, suggesting participation of a hydride ion (*19*). The participation in the reduction of the 5′-methylene group of 5′-deoxyadenosylcobalamin has been implicated by the observation that a transfer of ³H from 5′-deoxyadenosylcobalamin-5′-³H to water occurs during the reduction; this exchange also takes place in the absence of substrate and has an absolute dithiol requirement. Hogenkamp has discussed this exchange in terms of a postulated reaction sequence for the reduction of ribonucleoside triphosphates (*26*):

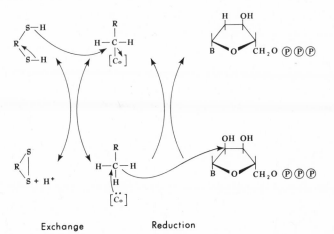

Exchange   Reduction

The tritium exchange may be rationalized in terms of this mechanism and in the knowledge that sulfhydryl hydrogens readily exchange with water hydrogens. It may be noted that ^{3}H from 5'-deoxyadenosylcobalamin-5'-^{3}H does not appear in the reduction product, evidently because the sulfhydryl–water hydrogen exchange is more rapid than the reduction.

The ribonucleoside triphosphate reductase of *L. leichmannii* is an allosteric enzyme, the activity of which is modified in a complex manner by deoxyribonucleoside triphosphates (*28, 29*). Reduction of each of the four substrates is maximally stimulated by a particular deoxyribonucleoside triphosphate, which Beck (*29*) terms a "prime effector." The data of Table 16–III illustrate the specific nature of the effector stimulation; prime effectors are indicated by the italicized data. The effector-induced stimulation of reductase activity appears to be countered in particular ways by deoxyribonucleoside triphosphates; for example, dTTP inhibits the dATP-activated reduction of CTP. It has been speculated that these complicated positive and negative allosteric effects produced by nucleotides may constitute a mechanism for ensuring that surpluses or shortages in the production of deoxyribonucleotides do not occur in the cell (*29*).

Thus, the *L. leichmannii* reductase, like that of *E. coli*, is an allosteric enzyme and it would appear that the specificity of the enzyme toward its ribonucleoside triphosphate substrates is determined by the particular nucleotide effectors that occupy allosteric sites on the enzyme. The properties of the two bacterial reductases are compared in Table 16–IV (*23*).

TABLE 16–III

INFLUENCE OF VARIOUS DEOXYRIBONUCLEOSIDE TRIPHOSPHATES ON
REDUCTION OF RIBONUCLEOSIDE TRIPHOSPHATES[a,b]

	Effector nucleotide added				
Substrate	None	dATP	dCTP	dGTP	dTTP
CTP	0.21	*6.95*	0.51	0.80	0.76
UTP	0.25	0.40	*2.32*	0.48	0.54
ATP	2.48	1.51	0.80	*7.00*	0.73
GTP	2.91	3.30	3.35	1.42	*7.01*

[a] From (*29*). Reproduced with permission.

[b] Components of the incubation mixtures were present in the following concentrations (mM): substrates (2.6), effectors (0.4), Mg^{2+} (16), dihydrolipoate (30), and 5,6-dimethylbenzimidazolylcobamide coenzyme (0.004). The data are in millimicromoles of reduction product formed per 20 minutes per 0.64 μg of purified ribonucleoside triphosphate reductase from *L. leichmannii*.

TABLE 16–IV

PROPERTIES OF RIBONUCLEOTIDE REDUCTASES FROM *Escherichia coli*
AND *Lactobacillus leichmannii*[a]

| Property | Ribonucleotide reductase | |
	E. coli	*L. leichmannii*
Substrates	Ribonucleoside *di*phosphates	Ribonucleoside *tri*phosphates
Products	2′-Deoxy NDP	2′-Deoxy NTP
Reductant	Thioredoxin	Thioredoxin
Mg^{2+}	Absolute requirement	Partial requirement (effector)
Cofactors	Nonheme iron	Coenzyme B_{12}
Hydroxyurea	Inhibitory	No effect
Molecular weight	$\sim$250,000	70,000–110,000
Subunits	Two nonidentical	Not known
Regulation	Allosterically regulated by nucleoside triphosphates. Repressible	Allosterically (?) regulated by nucleoside triphosphates. Repressible

[a] From (*23*). Reproduced with permission.

B. THE SYSTEM IN ANIMAL CELLS

The reduction of ribonucleotides by enzymes from animal cells was first demonstrated by Reichard and co-workers with crude preparations from chick embryo; the four common ribonucleotides were reduced and evidence for modulation of activity by nucleotides was obtained (*30*). The Novikoff hepatoma of the rat has been the source of a partly purified preparation of ribonucleotide reductase that has enabled deoxyribonucleotide synthesis to be studied in considerable detail; the work of Moore and colleagues has shown that this process has many similarities to that in *E. coli* (*31, 32*).

The partly purified reductase from the Novikoff tumor is specific for ribonucleoside diphosphates, and a single enzyme reduces CDP, UDP, ADP, and GDP. In analogy with the *E. coli* enzyme, the tumor reductase appears to require nonheme iron, as indicated by stimulation of reductase activity with iron salts and inhibition with iron-chelating agents. The purified reductase requires either a reductant such as dihydrolipoate or dithiothreitol, or an NADP-linked hydrogen transport system; the *E. coli* thioredoxin–thioredoxin reductase couple will link the tumor reductase with NADPH. There appears to be a thioredoxin-like hydrogen transport system in the tumor and in rat liver (*32*) that links nucleotide reduction with NADPH.

As with the bacterial reductases, a complex pattern of activation and
inhibition by nucleoside triphosphates has been demonstrated for the tumor reductase; dATP inhibits reduction of all four substrates (*32*). A
similar pattern of nucleotide regulatory effects was found with partly
purified reductase from rat embryo extracts (*33*). Thus, presuming that
analogy with the bacterial ribonucleotide reductases is valid, it would
appear that the animal reductases are allosteric enzymes subject to a complicated regulation by nucleotides; again, the function of such regulation
would seem to be that of ensuring a balanced supply of deoxyribonucleotides for DNA synthesis.

Ribonucleotide reductase activity has been related to proliferation of
animal cells in several ways. Turner *et al.* (*34*) found that the ribonucleotide
reductase activity present in cultured mouse fibroblasts (L cells) was
related to the mitotic cycle, being detectable only in the phase of DNA
synthesis. The reductase is scarcely detectable in adult liver; however, it
is present in tumors and in proliferating tissues and, as Elford *et al.* (*16*),
have reported, an excellent correlation exists between reductase activity
and growth rate in a series of rat hepatomas. Evidently, the activity of
ribonucleotide reductase varies in response to the cellular demand for DNA
matériel.

IV. Summary

The pathways of deoxyribonucleotide synthesis from ribonucleotides
are summarized below. The *trivial names* of the enzymes of ribonucleotide
reduction are as follows: ribonucleoside diphosphate reductase; ribonucleoside triphosphate reductase; thioredoxin reductase. Enzyme Commission numbers and systematic names have not yet been assigned.

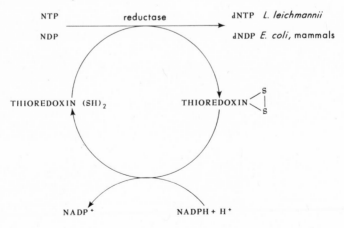

References

1. Hammarsten, E., Reichard, P., and Saluste, E., *J. Biol. Chem.* **183,** 105 (1950).
2. Rose, I. A., and Schweigert, B. S., *J. Biol. Chem.* **202,** 635 (1953).
3. Thomson, R. Y., Scotto, G. T., and Brown, G. B., *J. Biol. Chem.* **237,** 3510 (1962).
4. Larsson, A., and Neilands, J. B., *Biochem. Biophys. Res. Commun.* **25,** 222 (1966).
5. Grossman, L., and Hawkins, G. R., *Biochim. Biophys. Acta* **26,** 657 (1957).
6. Larsson, A., and Reichard, P., *Prog. Nucl. Acid Res. Mol. Biol.* **7,** 303 (1967).
7. Reichard, P., "The Biosynthesis of Deoxyribose," p. 13. Wiley, New York, 1967.
8. Brown, N. C., and Reichard, P., *J. Mol. Biol.* **46,** 25 (1969).
9. Brown, N. C., and Reichard, P., *J. Mol. Biol.* **46,** 39 (1969).
10. Holmgren, A., Reichard, P., and Thelander, L., *Proc. Nat. Acad. Sci. U.S.* **54,** 830 (1965).
11. Biswas, C., Hardy, J., and Beck, W. S., *J. Biol. Chem.* **240,** 3631 (1965).
12. Stryer, L., Holmgren, A., and Reichard, P., *Biochemistry* **6,** 1016 (1967).
13. Thelander, L., *Eur. J. Biochem.* **4,** 407 (1968).
14. Thelander, L., *J. Biol. Chem.* **245,** 6026 (1970).
15. Berglund, O., *J. Biol. Chem.* **244,** 6306 (1969).
16. Elford, H. L., Freese, M., Possamani, E., and Morris, H. P., *J. Biol. Chem.* **245,** 5228 (1970).
17. Larsson, A., *Biochemistry* **4,** 1984 (1965).
18. Durham, L. J., Larsson, A., and Reichard, P., *Eur. J. Biochem.* **1,** 92 (1967).
19. Batterham, T. J., Ghambeer, R. K., Blakely, R. L., and Brownson, C., *Biochemistry* **6,** 1203 (1967).
20. Follmann, H., and Hogenkamp, H. P. C., *Biochemistry* **8,** 4372 (1969).
21. Larsson, A., and Reichard, P., *J. Biol. Chem.* **241,** 2533 (1966).
22. Larsson, A., and Reichard, P., *J. Biol. Chem.* **241,** 2540 (1966).
23. O'Donovan, G. A., and Neuhard, J., *Bacteriol. Rev.* **34,** 278 (1970).
24. Blakley, R. L., and Barker, H. A., *Biochem. Biophys. Res. Commun.* **16,** 391 (1954).
25. Beck, W. S., and Hardy, J., *Proc. Nat. Acad. Sci. U.S.* **54,** 286 (1965).
26. Hogenkamp, H. P. C., *Annu. Rev. Biochem.* **37,** 225 (1968).
27. Orr, M. D., and Vitols, E., *Biochem. Biophys. Res. Commun.* **25,** 109 (1966).
28. Vitols, E., Brownson, C., Gardiner, W., and Blakely, R. L., *J. Biol. Chem.* **242,** 3035 (1967).
29. Beck, W. S., *J. Biol. Chem.* **242,** 3148 (1967).
30. Reichard, P., Canellakis, Z. N., and Canellakis, E. S., *J. Biol. Chem.* **236,** 2514 (1961).
31. Moore, E. C. and Reichard P. *J. Biol. Chem.* **239,** 3453 (1964).
32. Moore, E. C. and Hurlbert R. B. *J. Biol. Chem.* **241,** 4802 (1966).
33. Murphree S., Moore, E. C., and Beall, P. T., *Cancer Res.* **28,** 860 (1968).
34. Turner, M. K., Abrams, R., and Lieberman, I., *J. Biol. Chem.* **243,** 3725 (1968).

APPENDIX

METABOLISM AND BIOCHEMICAL EFFECTS OF PURINE AND PYRIMIDINE ANTIMETABOLITES

This Appendix summarizes the information given in the text regarding the metabolism and biochemical effects of purine and pyrimidine antimetabolites.

Compound	Structure	Reactions in which the compound and its derivatives participate		
8-Azaguanine		Base:	substrate for	hypoxanthine–guanine phosphoribosyl-transferase
				guanine deaminase
		Ribonucleoside monophosphate:	substrate for	guanylate kinase
6-Mercaptopurine		Base:	substrate for	hypoxanthine–guanine phosphoribosyl-transferase
		Ribonucleoside:	substrate for	purine nucleoside phosphorylase
		Ribonucleoside monophosphate:	inhibits	purine biosynthesis *de novo*
				IMP dehydrogenase
				GMP reductase
				adenylosuccinate synthetase
				adenylosuccinate lyase
6-Thioguanine		Base:	substrate for	hypoxanthine–guanine phosphoribosyl-transferase
		Ribonucleoside:	substrate for	purine nucleoside phosphorylase
		Ribonucleoside monophosphate:	inhibits	purine biosynthesis *de novo*
				IMP dehydrogenase
				GMP reductase

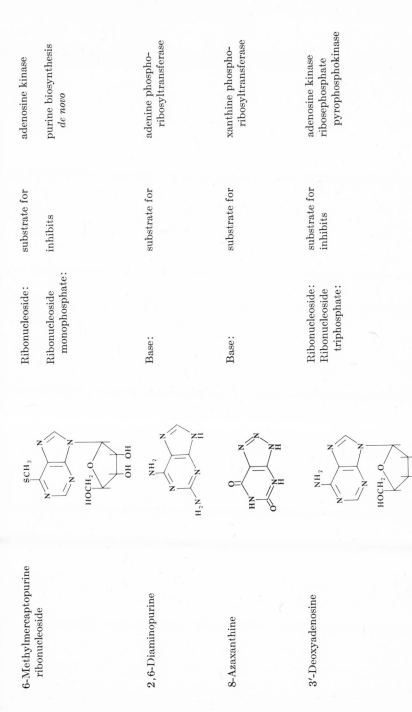

6-Methylmercaptopurine ribonucleoside

Ribonucleoside: substrate for adenosine kinase

Ribonucleoside monophosphate: inhibits purine biosynthesis *de novo*

2,6-Diaminopurine

Base: substrate for adenine phospho-ribosyltransferase

8-Azaxanthine

Base: substrate for xanthine phospho-ribosyltransferase

3'-Deoxyadenosine

Ribonucleoside: substrate for adenosine kinase

Ribonucleoside triphosphate: inhibits ribosephosphate pyrophosphokinase

Compound	Structure	Reactions in which the compound and its derivatives participate	
Formycin		Ribonucleoside triphosphate:	inhibits	ribosephosphate pyrophosphokinase
Psicofuranine		Nucleoside:	inhibits	ribosephosphate pyrophosphokinase GMP synthetase
Decoyinine		Nucleoside:	inhibits	ribosephosphate pyrophosphokinase

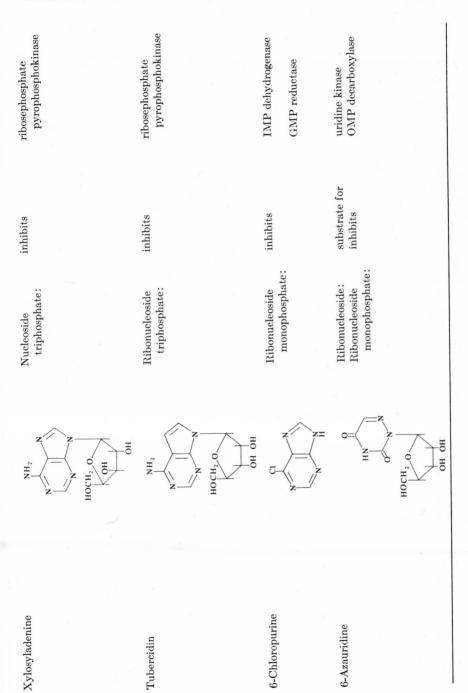

Xylosyladenine   Nucleoside
triphosphate: inhibits   ribosephosphate
pyrophosphokinase

Tubercidin   Ribonucleoside
triphosphate: inhibits   ribosephosphate
pyrophosphokinase

6-Chloropurine   Ribonucleoside
monophosphate: inhibits   IMP dehydrogenase
GMP reductase

6-Azauridine   Ribonucleoside:
Ribonucleoside
monophosphate: substrate for
inhibits   uridine kinase
OMP decarboxylase

Compound	Structure	Reactions in which the compound and its derivatives participate	
Deoxyxylopyranosylthymine		Nucleoside: inhibits	uridine phosphorylase
Deoxyglucopyranosylthymine		Nucleoside: inhibits	uridine phosphorylase
5-Fluorocytidine		Ribonucleoside: substrate for Deoxyribonucleoside: substrate for	uridine kinase cytidine deaminase deoxycytidine kinase
Cytosine arabinoside		Arabinoside: substrate for Triphosphate: inhibits	cytidine deaminase deoxycytidine kinase DNA nucleotidyl-transferase

5-Fluorouracil

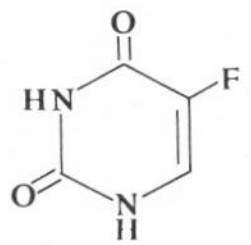

Base:	substrate for	uracil phospho-riboyltransferase
		dihydrouracil dehydrogenase
Ribonucleoside:	substrate for	uridine kinase
Deoxyribo-nucleoside:	substrate for	thymidine kinase
Deoxyribonucleo-side monophos-phate:	inhibits	thymidylate synthetase

5-Iodouracil

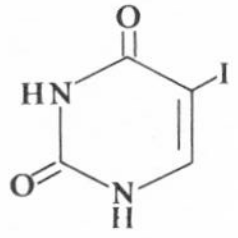

Base:	substrate for	dihydrouracil dehydrogenase
		uracil phospho-riboyltransferase
Deoxyribo-nucleoside:	substrate for	thymidine phospho-rylase
		thymidine kinase

5-Bromouracil

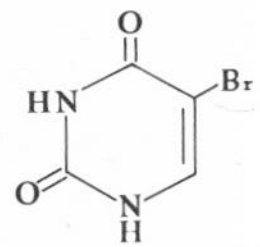

Base:	substrate for	dihydrouracil dehydrogenase
		uracil phospho-riboyltransferase
Deoxyribo-nucleoside	substrate for	thymidine phosphorylase
		thymidine kinase

AUTHOR INDEX

Numbers in parentheses are reference numbers and indicate that an author's work is referred to although his name is not cited in the text. Numbers in italics show the page on which the complete reference is listed.

A

Abrams, R., 137, 145 (4), 262 (34), *150,151*
Adams, W. S., 168 (43), *170*, 203 (7), *204*
Adelman, R. C., 60 (5), 61 (5), *67*
Adye, J. C., 127 (35), *134*
Ahmad, F., 109 (26), *122*
Albrecht, A. M., 119 (58), *123*
Aleman, V., 177 (12), *188*
Allan, P. W., 131 (62), *131*
Allison, A. J., 76 (8), *80*, 108 (24), *122*
Anderson, E. P., 63, *68*
Anderson, J. H., 145 (25), 147 (25), *151*
Anderson, P. M., 183, *188*
Appel, S. H., 182 (37), *189*
Armstrong, D. J., 52 (22), *56*
Arsenis, C., 153 (6), *169*
Ashton, D. M., 105 (12), *122*
Atkinson, D. E., 55 (33, 34), *56*, 90, *93*
Atkinson, M. R., 126 (13), 127 (25), 128 (25), *134*, 149 (51), *151*
Ayengar, P., 64 (20), *68*

B

Baer, H. P., 53 (26), *56*, 154 (8), 155 (12), 158, *169*

Bakay, B., 129 (45), *135*
Baker, F. A., 176 (10), *188*
Balis, M. E., 92, *93*, 119 (60), 120 (60), *123*, 127 (23), 129 (44), 130 (53), *134*, *135*, 148 (41), 153 (3), *151*, *169*
Bamman, B., 144 (24), *151*
Bandurski, R. S., 38(9), *56*
Barclay, M., 196 (24), *199*
Barclay, R. K., 196 (24), *199*, 210 (9), 213 (9), 214 (9), *224*
Barker, G. R., 87 (17), *92*
Barker, H. A., 258, *263*
Barner, H. D., 193 (14), *199*, 229, 232 (13), *242*
Bartlett, G., 19, *19*, 20 (39), *21*
Bartlett, G. R., 88 (29), *92*
Baserga, R., 222 (45), *225*
Bass, L. W., 6 (4), *20*
Batterham, T. J., 253 (19), 259 (19), *263*
Bayer, M., 145 (27), *151*
Beall, P. T., 262 (33), *263*
Beaven, G. H., 5 (1), 17 (1), *20*
Beavo, J. A., 10 (26), *21*
Beck, W. S., 215, *225*, 249, 258 (25), 260, *263*
Beiber, S., 128 (40), *135*

273

SUBJECT INDEX

A